T0189475

Wireless Networks

Series Editor

Xuemin Sherman Shen, University of Waterloo, Waterloo, ON, Canada

The purpose of Springer's Wireless Networks book series is to establish the state of the art and set the course for future research and development in wireless communication networks. The scope of this series includes not only all aspects of wireless networks (including cellular networks, WiFi, sensor networks, and vehicular networks), but related areas such as cloud computing and big data. The series serves as a central source of references for wireless networks research and development. It aims to publish thorough and cohesive overviews on specific topics in wireless networks, as well as works that are larger in scope than survey articles and that contain more detailed background information. The series also provides coverage of advanced and timely topics worthy of monographs, contributed volumes, textbooks and handbooks.

** Indexing: Wireless Networks is indexed in EBSCO databases and DPLB **

Ying Chen • Ning Zhang • Yuan Wu •
Sherman Shen

Energy Efficient Computation Offloading in Mobile Edge Computing

Ying Chen
Computer School
Beijing Information Science and
Technology University
Beijing, China

Ning Zhang
Department of Electrical and Computer
Engineering
University of Windsor
Windsor, ON, Canada

Yuan Wu
State Key Lab of Internet of Things for
Smart City
University of Macau
Taipa, China

Sherman Shen
Department of Electrical and Computer
Engineering
University of Waterloo
Waterloo, ON, Canada

ISSN 2366-1186 ISSN 2366-1445 (electronic)
Wireless Networks
ISBN 978-3-031-16824-6 ISBN 978-3-031-16822-2 (eBook)
https://doi.org/10.1007/978-3-031-16822-2

This Springer imprint is published by the registered company Springer Nature Switzerland AG
The registered company address is: Gewerbestrasse 11, 6330 Cham, Switzerland

Preface

With the proliferation of mobile devices and development of Internet of Things (IoT), more and more computation-intensive and delay-sensitive applications are running on terminal devices, which result in high-energy consumption and heavy computation load of devices. Due to the size and hardware constraints, the battery lifetime and computing capacity of terminal devices are limited. Consequently, it is hard to process all of the tasks locally while satisfying Quality of Service (QoS) requirements for devices. Mobile Cloud Computing (MCC) is a potential technology to solve the problem, where terminal devices can alleviate the operating load by offloading tasks to the cloud with abundant computing resources for processing. However, as cloud servers are generally located far away from terminal devices, data transmission from terminal devices to cloud servers would incur a large amount of energy consumption and transmission delay. Mobile Edge Computing (MEC) is considered as a promising paradigm that deploys computing resources at the network edge in proximity of terminal devices. With the help of MEC, terminal devices can achieve better computing performance and battery lifetime while ensuring QoS. The introduction of MEC also brings the challenges of computation offloading and resources management under energy-constrained and dynamic channel conditions. It is of importance to design energy-efficient computation offloading strategies while considering the dynamics of task arrival and system environments.

In this book, we provide a comprehensive review and in-depth discussion of the state-of-the-art research literature and propose energy-efficient computation offloading and resources management for MEC, covering task offloading, channel allocation, frequency scaling, and resource scheduling. In Chap. 1, we provide a comprehensive review on the background of MCC, MEC, and computation offloading. Then, we present the characteristics and typical applications of MCC and MEC. Finally, we summarize the challenges of computation offloading in MEC. In Chap. 2, we propose an Energy Efficient Dynamic Computing Offloading (EEDCO) scheme to minimize energy consumption and guarantee terminal devices' delay performance. In Chap. 3, to further improve energy efficiency combined with tail energy, we propose a Computation Offloading and Frequency Scaling for Energy

Efficiency (COFSEE) scheme to jointly deal with the stochastic task allocation and CPU-cycle frequency scaling to achieve the minimum energy consumption while guaranteeing the system stability. In Chap. 4, we investigate delay-aware and energy-efficient computation offloading in a dynamic MEC system with multiple edge servers, and an end-to-end Deep Reinforcement Learning (DRL) approach is presented to select the best edge server for offloading and allocate the optimal computational resource such that the expected long-term utility is maximized. In Chap. 5, we study the multi-task computation offloading in multi-access MEC via non-orthogonal multiple access (NOMA), and accounting for the time-varying channel conditions between the ST and edge-computing servers, an online algorithm, which is based on DRL, is proposed to efficiently learn the near-optimal offloading solutions. In Chap. 6, we conclude the book and give directions for future research.

We believe that the presented computation offloading and energy management solutions and the corresponding research results in this book can provide some valuable insights for practical applications of MEC and motivate new ideas for future MEC-enabled IoT networks.

We would like to thank Mr. Kaixin Li and Mr. Yongchao Zhang from Beijing Information Science and Technology for their contributions to this book. We also would like to thank all members of the BBCR group at the University of Waterloo for their valuable discussions and insightful suggestions. Special thanks go to Susan Lagerstrom-Fife and Shina Harshavardhan from Springer Nature for their help throughout the publication process.

Beijing, China — Ying Chen
Windsor, ON, Canada — Ning Zhang
Macao, China — Yuan Wu
Waterloo, ON, Canada — Sherman Shen

Acknowledgements

This work was supported in part by the National Natural Science Foundation of China under Grants 61902029 and 62072490, the Scientific Research Project of Beijing Municipal Education Commission under Grant KM202011232015, in part by the Joint Scientific Research Project Funding Scheme between Macao Science and Technology Development Fund and the Ministry of Science and Technology of the People's Republic of China under Grant 0066/2019/AMJ, in part by the Macao Science and Technology Development Fund under Grants 0060/2019/A1 and 0162/2019/A3, in part by FDCT SKL-IOTSC(UM)-2021-2023, in part by the Guangdong Basic and Applied Basic Research Foundation (2022A1515011287), and in part by the Natural Sciences and Engineering Research Council of Canada.

Contents

1 Introduction 1
1.1 Background 1
1.1.1 Mobile Cloud Computing 2
1.1.2 Mobile Edge Computing 8
1.1.3 Computation Offloading 12
1.2 Challenges 17
1.3 Contributions 19
1.4 Book Outline 21
References 21

2 Dynamic Computation Offloading for Energy Efficiency in Mobile Edge Computing 27
2.1 System Model and Problem Statement 27
2.1.1 Network Model 28
2.1.2 Task Offloading Model 28
2.1.3 Task Queuing Model 30
2.1.4 Energy Consumption Model 30
2.1.5 Problem Statement 31
2.2 EEDCO: Energy Efficient Dynamic Computing Offloading for Mobile Edge Computing 32
2.2.1 Joint Optimization of Energy and Queue 32
2.2.2 Dynamic Computation Offloading for Mobile Edge Computing 34
2.2.3 Trade-Off Between Queue Backlog and Energy Efficiency 36
2.2.4 Convergence and Complexity Analysis 37
2.3 Performance Evaluation 40
2.3.1 Impacts of Parameters 40
2.3.2 Performance Comparison with EA and QW Schemes 48
2.4 Literature Review 50
2.5 Summary 57
References 57

3 Energy Efficient Offloading and Frequency Scaling for Internet of Things Devices 61
3.1 System Model and Problem Formulation 61
3.1.1 Network Model 63
3.1.2 Task Model 63
3.1.3 Queuing Model 65
3.1.4 Energy Consumption Model 65
3.1.5 Problem Formulation 67
3.2 COFSEE: Computation Offloading and Frequency Scaling for Energy Efficiency of Internet of Things Devices 68
3.2.1 Problem Transformation 68
3.2.2 Optimal Frequency Scaling 71
3.2.3 Local Computation Allocation 72
3.2.4 MEC Computation Allocation 74
3.2.5 Theoretical Analysis 75
3.3 Performance Evaluation 79
3.3.1 Impacts of System Parameters 80
3.3.2 Performance Comparison with RLE, RME and TS Schemes 84
3.4 Literature Review 87
3.5 Summary 91
References 92

4 Deep Reinforcement Learning for Delay-Aware and Energy-Efficient Computation Offloading 97
4.1 System Model and Problem Formulation 97
4.1.1 System Model 97
4.1.2 Problem Formulation 99
4.2 Proposed DRL Method 103
4.2.1 Data Prepossessing 104
4.2.2 DRL Model 105
4.2.3 Training 108
4.3 Performance Evaluation 111
4.4 Literature Review 117
4.5 Summary 120
References 120

5 Energy-Efficient Multi-Task Multi-Access Computation Offloading via NOMA 123
5.1 System Model and Problem Formulation 123
5.1.1 Motivation 123
5.1.2 System Model 125
5.1.3 Problem Formulation 126
5.2 LEEMMO: Layered Energy-Efficient Multi-Task Multi-Access Algorithm 127
5.2.1 Layered Decomposition of Joint Optimization Problem 128

5.2.2 Proposed Subroutine for Solving Problem (TEM-E-Sub) 129
5.2.3 A Layered Algorithm for Solving Problem (TEM-E-Top) ... 131
5.2.4 DRL-Based Online Algorithm ... 133
5.3 Performance Evaluation ... 136
5.3.1 Impacts of Parameters ... 136
5.3.2 Performance Comparison with FDMA Based Offloading Schemes ... 140
5.4 Literature Review ... 143
5.5 Summary ... 147
References ... 148

6 Conclusion ... 153
6.1 Concluding Remarks ... 153
6.2 Future Directions ... 155
References ... 156

Acronyms

3GPP	The Third-Generation Partnership Project
4G	The Fourth Generation Communication System
5G	The Fifth Generation Communication Technology
A3C	Asynchronous Advantage Actor-Critic
AR	Augmented Reality
AS	Additional-Sampling
BS	Base Station
C-RAN	Centralized RAN
CBS	Current Best Solution
CBV	Current Best Value
CPU	Central Processing Unit
CVX	Convex Optimization
DNN	Deep Neural Network
DP	Dynamic Programming
DRL	Deep Reinforcement Learning
DT	Data Transmission
E2E_DRL	End-to-End DRL
ECS	Edge-Computing Server
ETSI	European Telecommunications Standards Institute
ID	Idle
IIoT	Industrial Internet of Things
IoT	Internet of Things
IoV	Internet of Vehicles
IT	Information Technology
LFP	Linear-Fractional Programming
LP	Linear Programming
LSTM	Long Short-Term Memory
LTE	Long-Term Evolution
MA-MEC	Multi-Access Mobile Edge Computing
MCC	Mobile Cloud Computing
MDP	Markov Decision Process

MEC	Mobile Edge Computing
MEC ENV	MEC environment
MSE	Mean Square Error
NOMA	Non-Orthogonal Multiple Access
NP	Non-deterministic Polynomial
PC	Personal Computer
QoS	Quality of Service
RAN	Radio Access Network
RL	Reinforcement Learning
RLE	Round-robin Local Execution
RME	Round-robin MEC Execution
RRC	Radio Resource Control
SDN	Software Defined Network
SIC	Successive Interference Cancellation
TA	Tail
TEM	Total Energy Minimization
UAV	Unmanned Aerial Vehicle
VR	Virtual Reality
WAN	Wide Area Network
WD	Wireless Device
WiFi	Wireless Fidelity

Chapter 1
Introduction

1.1 Background

The development of cellular mobile communication has greatly changed our way of life. Since the emergence of the first generation of communication technologies, a new generation of mobile communication systems appears about every ten years. The emergence of each new generation of mobile communication systems brings new experiences to people. The emergence and development of the fifth generation communication technology (5G) has brought a new revolution to mobile communication technology [1, 2]. 5G has higher spectral efficiency, can support more users, and ensure user experience while providing higher communication rates. Many new application technologies also emerge with the development of 5G, such as immersive entertainment, car networking and augmented reality/virtual reality.

With the development of 5G technology, mobile computing technology is also developing rapidly [3]. The development of mobile technology has also promoted the development of mobile terminal equipment. In recent years, with the development of communication technology, the number of mobile devices is also growing rapidly. Mobile devices have become an integral part of people's life and work. In order to better meet the needs of people's production and life, more and more computation-intensive applications have appeared. The increasingly complex mobile applications such as mobile payment, mobile medical care, mobile games and virtual reality have put forward higher requirements on the computing power, storage capacity, battery capacity and security of mobile devices. However, due to the limitations of the weight, size and battery capacity, it is impossible for mobile devices to independently handle these tasks locally, which will degrade the performance of mobile devices and seriously affect the application experience them. Therefore, in order to provide a better experience to mobile users, a platform is needed to help mobile devices handle these computation-intensive tasks to relieve the pressure on mobile devices. Mobile cloud computing (MCC) is introduced as

Y. Chen et al., *Energy Efficient Computation Offloading in Mobile Edge Computing*, Wireless Networks, https://doi.org/10.1007/978-3-031-16822-2_1

a platform with powerful computing power and abundant computing resources to help mobile devices handle complex tasks [4].

MCC is the combination of cloud computing, storage and other resources to provide a better application experience for mobile devices. Mobile devices offload computation-intensive tasks to the cloud. But cloud computing centers are often located far from mobile devices. Therefore, offloading a large number of tasks to the cloud for processing will introduce huge delays and affect the user experience. At the same time, because too much data is offloaded, the network will become congested.

Therefore, mobile edge computing (MEC) is introduced as an emerging computing mode [5]. MEC is a distributed computing method based on mobile communication networks, which extends the functions of cloud computing by performing computations at the edge of the network. MEC can be seen as a small-scale cloud server running at the edge of the mobile network. By shifting the load of cloud computing to edge servers, MEC helps reduce congestion on mobile networks and reduce latency, thereby improving the quality of experience for mobile users. MEC helps save costs and reduce latency [6]. With MEC, mobile users can get better experience.

Although MEC can make data communication more convenient, it also faces some challenges [7]. Since the tasks generation process on mobile devices are stochastic and dynamic, we cannot precisely predict the amount of computing resources allocated to mobile devices based on the size of the amount of task. At the same time, during the communication process, since the state of the wireless channel also changes dynamically, we cannot precisely predict the quality of the network state at every moment. Since many computation-intensive tasks are delay-sensitive, it is necessary to find a way to control the bound of latency. In order to ensure the battery life of mobile devices, we also pay attention to the consumption of energy. If the energy consumption of the mobile device is too large, it will affect the usage time of the mobile device and reduce the user experience. Therefore, how to design an energy-efficient task offloading scheme is crucial.

In this chapter, we introduce the background of MCC, MEC, and computation offloading. At the same time, we summarize our contributions. Specifically, in Sect. 1.1, we present the background. In Sect. 1.2, we present the challenges encountered in computation offloading in MEC. In Sect. 1.3, we summarize our contributions. In Sect. 1.4, we give the outline of the book.

1.1.1 Mobile Cloud Computing

With the development of mobile computing and the wide deployment of mobile applications, mobile devices have become an indispensable part of people's production and life. Due to the powerful computing power and abundant computing resources, MCC provides mobile users with a better application experience. The definition of MCC is generally the use and delivery mode in which mobile devices obtain the required infrastructure, platform, software and other resources or information services from the cloud through wireless networks in an on-demand and easily scalable manner [4]. MCC is a combination of mobile computing and cloud computing. Cloud computing is a technology that can provide resources and services through a network, including computing power, storage resources, infrastructure, platforms, applications and some other requirements. Fortunately, cloud computing can provide dynamic resources for computing, storage and services, which will overcome the limitations in smart mobile devices [8].

Cloud computing is the core of MCC [9]. The main service models of MCC are mobile network as a service, mobile cloud infrastructure as a service, mobile data as a service, mobile platform as a service and mobile software as a service [10]. Similar to traditional cloud computing deployment models (private cloud, public cloud, community cloud and hybrid cloud) [9], the deployment modes of MCC include temporary mobile cloud, private mobile cloud, public mobile cloud, mobile community cloud and enterprise mobile cloud [10].

In MCC, mobile users connect to the cloud through wireless access points such as base stations, and then obtain services such as computing and storage from the cloud. Content providers can place resources such as videos and files on the cloud to provide users with richer services. MCC is also known as 3+ combined technology. MCC is a computation strategy that combines mobile computing, cloud computing and Internet technologies to provide users with a new platform for using IT services [11].

1.1.1.1 Architecture of Mobile Cloud Computing

One of the main goals of cloud computing is to make it easier for small business and individual users to use cloud resources and services. Similarly, one of the goals of MCC is to provide users with an enhanced user experience, such as extending the battery life of the device, increasing the speed at which the device can process tasks, and providing more comprehensive services. Compared with cloud computing, MCC faces different problems and challenges, such as network connection, communication load, network bandwidth and so on.

Currently, in the MCC architecture, mobile devices can access mobile services in two ways. The two ways are access via mobile telecommunication network or via wireless network respectively [4, 12]. When using the former method to access mobile services, mobile devices need to rely on communication base stations or satellites. When using the latter method to access mobile services, the mobile device needs to be connected to a WiFi access point. But no matter which method is used, access to the Internet is required to obtain services and resources in the cloud. However, in comparison, using WiFi access will have better performance in terms of latency and energy consumption.

1.1.1.2 Characteristics of Mobile Cloud Computing

MCC combines the characteristics of mobile computing and cloud computing [13]. The characteristics of MCC are as follows:

Mobility The advantage of mobile computing is mobility. When a mobile device requests resource, the user does not need to move to a specific location in advance, that is, the user can obtain the required resource anywhere. The biggest characteristic of mobile users is that the user's location is in the process of constant change. Therefore, it is very important to locate the location of mobile users accurately and efficiently. Because only by accurately grasping the user's location, can the user be provided with the required services anytime, anywhere. MCC is designed based on the model of cloud computing. Therefore, MCC can be used to collect location information of mobile devices, so as to be ready to serve mobile users at any time.

Communication MCC can realize the communication between beings, such as social software. This feature brings people closer together.

Entertainment With the development of communication technology and the needs of people's lives, more and more applications bring happiness to people. People can use smart mobile devices to watch movies, listen to music and engage in other entertainment activities.

Cloud Computing Cloud computing centers have a large number of computing resources and services. Mobile users can get the resources they need from the cloud at any time, which is convenient and fast.

On-Demand Self-Service The needs of every mobile user are different. If the service is customized for each user according to the personality, the service load will increase. The cloud technology can realize the sharing of resources among different users, so as to reduce the service cost.

Resource Pool A collection of various hardware and software involved in a cloud computing data center. The types of hidden devices can be divided into computing resources, storage resources and network resources.

Elastic Service Mobile providers can provide services flexibly according to the needs of users.

Diversity of Network Access Users can use mobile phones, PCs or workstations to access the cloud.

Weakening the Hardware Limitations of Terminal Devices It is difficult to process computation-intensive data solely on mobile terminal devices. Because cloud computing performs computing and data storage based on mobile networks, the computing power of mobile devices themselves is ignored. Relying on cloud computing can effectively solve the hardware limitations of mobile devices.

Convenient Data Storage When MCC stores data, it is all carried out in the mobile Internet, which not only facilitates users to store data, but also provides users with a certain data storage space.

Computing Migration Computing migration is the core technology of MCC, which is mainly used to solve the problem of limited computing and storage resources of mobile terminal devices. Mobile devices can migrate computing tasks to cloud data centers or cloud servers for execution. In this way, the burden of the mobile device can be reduced, which can not only prolong the use time of the mobile device, but also enable the user to obtain a better service experience. There are also scholars in academia who have studied this issue. Goyal et al. [14] studied the computing and storage capabilities of mobile devices, and proposed some computation migration schemes.

1.1.1.3 Cloudlet

In the basic model of MCC, different communication technologies (such as 4G, 5G and WiFi) are used to communicate between mobile devices in the network. When communicating, the mobile device first determines whether the user is a legitimate user. Then, the user's service request is forwarded to the cloud server, and the cloud server processes the request and provides the required service.

However, due to the long distance between the cloud computing center and the mobile device, it will lead to large delay during data transmission. Therefore, to address the challenge of latency in MCC, Sat et al. [15] propose a new cloud architecture, namely Cloudlet. Cloudlet can provide cloud services to users remotely with less latency and greater throughput [16].

A Cloudlet is a trusted, resource-rich computer or cluster of computers that can be networked or disconnected, allowing nearby mobile devices to use it [15]. Cloudlet supports computation-intensive and interactive applications and provides powerful computing resources and lower latency for mobile devices. Logical proximity between Cloudlet and mobile devices is necessary. Cloudlets can be located in cellular base stations or WiFi access points.

Cloudlet overcomes the WAN latency problem and provides low-latency, high-bandwidth real-time interactive services through the local area network. Cloudlets can be further subdivided into mobile cloudlets [17] and fixed cloudlets [18]. Fixed Cloudlet mode provides cloud computing services with desktop computers with strong computing power, and usually this connection mode can provide larger bandwidth and computing resources. The mobile Cloudlet mode builds Cloudlet through mobile terminals, aiming to provide access services anytime and anywhere.

In the Cloudlet model, the mobile user is connected to the Cloudlet [19]. When a mobile user needs to request cloud resources, it first communicates with Cloudlet through WiFi technology, and then Cloudlet sends the request to the cloud center. The request result is returned to the mobile device through Cloudlet again, reducing latency and energy consumption.

Cloudlet is mainly used in the integration scenarios of mobile computing and cloud computing. Cloudlet is suitable for four scenarios, including highly responsive cloud services, scaling edge analytics, privacy policy enforcement, and shielding cloud outages.

1.1.1.4 Fog Computing

With the development of IT technology, more and more people and businesses are turning to smart devices to handle everyday tasks. These devices generate huge amounts of data every day. If all this data is transmitted to the cloud for processing, it will result in greater latency and energy consumption. Hence, the fog computing paradigm came into being. "Fog computing" was originally proposed by researchers at Cisco [20].

We know that fog is a cloud close to the ground. Therefore, the idea of fog computing is to extend the cloud computing architecture to the edge of the network [21]. Compared with cloud computing, the architecture adopted by fog computing is distributed and closer to the network edge. Unlike cloud computing, which stores almost all data and data processing on the cloud, fog computing concentrates data and data processing in devices at the edge of the network, making data storage and processing more dependent on local devices.

The architecture of fog computing mode has three layers, namely cloud data layer, fog layer and mobile device layer. In general cloud computing, a mobile terminal device is usually connected to a cloud server through a WAN [22]. However, in fog computing, mobile terminal devices are first connected to the fog layer through a local area network. In the fog layer, there are microservers (often called fog nodes) that can provide computing resources and storage services. Fog nodes are usually deployed in public places, such as shopping malls, parks, stations, etc. Fog nodes usually consist of some devices that provide networking capabilities, such as routers. Fog computing can handle delay-sensitive tasks from devices while reducing energy consumption and achieving efficient use of resources [23].

The characteristics of fog computing are as follows [24]:

Low Latency Fog devices are usually deployed at the edge of the network. Compared with remotely deployed cloud devices, fog devices can provide computing and storage services for terminal devices nearby. Therefore, the data transmission time is greatly shortened, the data processing speed is improved, and the communication pressure on the backbone network is reduced.

Distribution Cloud computing centers usually contain some high-performance server clusters, which are typical centralized models. Fog computing is a typical distributed model. Fog devices are widely distributed at the edge of each network, which can provide support and guarantee for those services with real-time and mobility requirements.

Perception Compared with the deployment of cloud devices, the deployment of fog devices is denser and more widely distributed. Fog equipment has the ability to perceive and obtain information about the surrounding environment at any time and quickly.

Mobility In a fog computing network, the deployment location of fog devices can be fixed or mobile. They can serve both fixed terminal equipment and mobile terminal equipment. Stationary fog devices can be gateways, routers, etc. deployed in fixed locations, and mobile fog devices can be moving cars, flying drones, and the like.

Low Energy Consumption Fog computing is more energy efficient than cloud computing. In the cloud computing architecture, all data is transmitted to the cloud data center for centralized processing, which will consume more energy. In the fog computing architecture, most of the data does not need to be sent to the cloud computing center, but only needs to be sent to the nearby fog devices for processing, which greatly saves the energy consumption of mobile devices.

Heterogeneity Fog devices can be composed of various computing devices or network devices that are relatively rich in computing resources and storage resources. These devices have great differences in function and performance, and rely on different communication protocols and transmission media. Based on the service-oriented concept, fog computing uses virtualization technology to integrate these heterogeneous devices into a virtualization platform that can provide unified storage, computation and services.

1.1.1.5 Data Security and Privacy Protection

The development of MCC has brought great convenience to people, but it also faces security threats such as data leakage and privacy exposure.

Safety for Mobile Users Although cloud services are available over the Internet, mobile devices can also be attacked [25]. As the variety of applications becomes more and more diverse, the possibility of being attacked by a virus when a user installs an application increases [13]. At this time, the user can install the matching security management software to protect the mobile device, which is also the most common way to protect the mobile device. When sharing data, ring signature technology based on public key encryption mechanism is generally used to achieve privacy protection, which can prevent cloud service providers from illegally obtaining user identity information [26]. To deal with this problem, Shabtai et al. [27] proposed a knowledge-based temporal abstraction method to integrate software installation events with the security knowledge base of mobile devices to detect possible applications.

Data Security MCC can enable users to store large amounts of data on the cloud, alleviating the requirements on the mobile device itself. However, storing data on the cloud faces the problem of data leakage. With the development of cloud computing, incidents about data leakage appear frequently, which has a negative impact on people's experience. Therefore, it is very important to ensure the security of data on the cloud. Khalil et al. [28] proposed an identity management system architecture that allowed users to access cloud services only if the token sent by the user was the same as the token the user has stored in the cloud.

1.1.1.6 Challenges of Mobile Cloud Computing

MCC indeed has many advantages. However, with the development of the IoT and the increase number of computation-intensive applications, there are still many challenges [29]. First, there is a lack of battery life when running large games and dealing with power-hungry applications like audio and video streaming. Second, MCC lacks an accepted standard for deploying services across multiple cloud computing service providers. Third, the access management of MCC is not comprehensive enough, such as mobile nodes accessing the cloud through different radio access technologies. Fourth, due to the limitations of processor speed and memory, MCC cannot efficiently run applications with huge computation offloading. Fifth, high network latency and low bandwidth are also issues that cannot be ignored in MCC.

Since the cloud computing center is generally located far away from the mobile device, a large transmission delay will occur when the mobile device offloads the task, which is intolerable for some computation-sensitive applications. At the same time, since the mobile device is far away from the cloud computing center, it consumes a lot of energy when transmitting data, which is very unfriendly to mobile devices with small size and low battery capacity. Therefore, we need a more suitable mobile computing method. Therefore, the emerging MEC has become a hotspot in mobile computing research area [30].

1.1.2 Mobile Edge Computing

In recent years, with the rapid development of technologies such as IoT, cloud computing and big data, the data generated by mobile devices has exploded. The traditional cloud computing model needs to upload massive data to the cloud server. However, because the cloud server is far away from the terminal device, problems such as response delay, network interference, and energy consumption will inevitably occur during the transmission process. Cloud computing has powerful computing power, it can solve the problem of a large number of task computing and mobile device battery consumption that cannot be solved by mobile devices alone [31]. However, with the development of smart terminal devices and the ever-changing demand for services, users are pursuing a more smooth experience. Therefore, the requirement for data transfer rate and low latency becomes higher. This is also the reason that cloud computing is difficult to meet the needs of many current scenarios [32].

With the development of 5G technology, data exchange between mobile devices and remote cloud servers may become more frequent. It is difficult to achieve millisecond-level communication delays only by relying on cloud computing mode for data transmission. Therefore, in order to solve the delay and energy consumption problems in the cloud computing mode, some scholars propose to migrate the functions of the cloud to the network edge. Therefore, MEC emerges as a new network structure and computing paradigm.

1.1.2.1 Definition of Mobile Edge Computing

MEC is an emerging mode that utilizes mobile base stations to extend cloud computing services to the network edge. MEC servers are mainly deployed in base stations near the network edge of network equipment. More and more emerging mobile applications benefit from offloading tasks from mobile devices to edge servers for execution [33]. MEC supports scenarios such as 3G/4G/5G, wired network and wireless network [34]. MEC can provide advantages for industrial, entertainment, personal computing and other applications [35].

1.1.2.2 Architecture of Mobile Edge Computing

The architecture of MEC is typically a three-layer architecture, namely cloud-edge-terminal three-layer architecture, which are cloud layer (cloud computing), edge layer (MEC server), and terminal layer (mobile terminal device) [36]. In MEC, the mobile terminal device does not communicate with the server directly, but communicates with the MEC server by base station or a wireless access point. The MEC server is deployed at the edge closer to the terminal device, which can provide computing and caching services for the mobile device, and can reduce the delay and energy consumption of the mobile device.

ETSI proposed the MEC architecture structure, composed of gasoline functional elements and reference points allowing their interactions [37]. The MEC architecture consists of mobile terminal equipment, mobile edge system layer management, mobile edge service layer management and mobile edge server. As the MEC architecture becomes more mature, the application of the MEC architecture to the IoT system can solve the internal attack problem that cannot be effectively solved by the security strategy in the traditional IoT system [38].

1.1.2.3 Advantages of Mobile Edge Computing

MEC is close-range, large-bandwidth, low-latency, localized computing. Because MEC can extend resources and services to the edge of the network, it can make up for the lack of security of data storage and high latency of services in cloud computing mode. There are significant differences between MEC and cloud computing in terms of distance and latency from mobile terminal devices. The advantages of MEC are lower latency, energy saving for mobile devices, and protection of mobile device privacy [39]. There are the following typical advantages.

1. By transmitting the data generated by the mobile device to the network edge for calculation, the back-to-city traffic of the cellular network is greatly reduced, which not only makes the calculation more efficient, but also reduces the transmission cost.
2. For some delay-sensitive applications, requesting services through MEC can greatly reduce the delay and improve the user experience quality.
3. MEC runs at the edge of the network and does not logically depend on other parts of the network, which is very important for applications with high security requirements.
4. MEC servers usually have high computing power, which is very suitable for some applications that require a lot of data processing.
5. MEC can protect privacy and reduce the risk of data leakage by processing, storing and discarding appropriate data.

1.1.2.4 Applications of Mobile Edge Computing

Because MEC is close to terminal equipment and has certain computing power and storage capacity, MEC has been applied in many fields. As a promising distributed computing paradigm, MEC can be applied in various scenarios, such as mobile, wireless and wired scenarios. MEC moves computing, storage and service resources to the edge of the network. MEC deploys edge servers at the edge of the network close to the client [40]. MEC can provide a large number of computing resources and storage resources for mobile devices to relieve the resource shortage of mobile devices. Since the MEC server is deployed at the network edge near the mobile device, the data can be processed in a timely manner, which greatly reduces the transmission delay and enables those delay-sensitive applications to obtain a better service experience.

(A) Smart City

In a large city with a large population, a large amount of data is generated all the time, and if these data are processed through the cloud center, it will cause a huge network burden. If these data are processed nearby, not only can the processing power be improved, but also the network load can be reduced. In large cities, there are many services that need to be processed in real time, such as public safety.

Therefore, smart cities are also getting more and more attention [41, 42]. Through edge computing, data transmission time can be reduced, thereby improving user experience. For some location-based applications, real-time processing and precise judgment are required, such as navigation. Mobile devices can hand over their location information to edge servers for processing, which can get quick responses.

(B) Smart Healthcare
MEC can also be applied in smart medical and healthcare work. Liu et al. [43] designed a food recognition system for meal evaluation based on edge computing, which offloaded the recognition task to a nearby edge server for processing, which can effectively solve the problems of latency and energy consumption.

In the new era, the traditional medical model is gradually transforming into a smart medical model, which can make full use of medical resources. In China's fight against the new crown pneumonia epidemic in 2020, many hospitals have established smart medical infrastructure. Nurses can upload patient information to the system, and then doctors can remotely treat them, which can effectively solve the problem of the tension of medical staff in special times. In healthcare, real-time monitoring and feedback are needed to avoid falls in stroke patients. Data analysis through edge computing can reduce latency and enable quick responses.

(C) Smart Grid
With the development of industrialization and the acceleration of urbanization, people's demand for power resources has become higher and higher. The management method of traditional power grid can not adapt to the current situation, and the smart grid emerges as the times require [44]. The smart grid can guarantee the intelligent monitoring and management of the whole process of electricity from the power plant to the user.

(D) Virtual and Augmented Reality
Virtual reality and augmented reality enhance people's perception of the world and change people's way of life by combining real world scenes with virtual worlds [45]. MEC provides a powerful help for the development of VR. By offloading the data generated by VR/AR to the edge server for execution, the problems of insufficient computing power, storage capacity and battery capacity of VR/AR devices can be alleviated. Utilizing the short-range transmission of MEC, the problem of high latency can be solved, and at the same time, the VR device can be freed from the limitation of the wired network.

1.1.2.5 Challenges of Mobile Edge Computing

As a core technology in 5G, MEC provides cloud computing capabilities in the radio access network, improving the quality of service for mobile devices and reducing core network load traffic. But with the advent of the era of big data, the scale of mobile devices is also expanding. The resulting mobile data traffic is also increasing, resulting in a sharp increase in the energy consumption of the system.

From the perspective of mobile devices, when processing computing tasks, they hope to save their own energy consumption and speed up computing tasks. However, the computing resources and computing power possessed by edge servers are also limited, and it is impossible to meet the needs of all mobile devices. Therefore, when offloading computing tasks to edge servers, it is necessary to consider the amount of offloading and the time of task offloading, so that the edge servers can fully handle these tasks to meet the needs of mobile devices. However, since computation offloading is a complex process affected by different factors. Therefore, it is impossible to find a completely general offloading strategy.

Because of this, the research on computation offloading problem is also put on the agenda. A good computation offloading method can not only save a lot of system energy consumption, but also reduce system delay and provide users with faster, more stable and efficient services [46].

1.1.3 Computation Offloading

With the development of IoT technology and the promotion of mobile devices, the traffic generated by mobile devices is also increasing rapidly. However, due to their limited resources and computing performance, smart mobile devices may face insufficient capacity when processing computation-sensitive and delay-sensitive applications. The edge computing model that uses edge servers to process data emerges and complements MCC. However, mobile devices are characterized by small size, weak computing power, and low battery capacity. Therefore, how to reasonably utilize the limited computing resources of edge servers has become an important issue that needs to be solved urgently in edge computing. Edge computing provides cloud computing services near mobile devices through wireless access points, which meets the needs of mobile devices for fast interactive response. How to rationally utilize the services provided by edge servers, how to offload computation-intensive tasks from mobile devices to edge servers, and how to perform efficient and reasonable offloading have become important research contents of edge computing.

Computation offloading is a key technology in edge computing, which can provide resource-constrained mobile devices with computing resources to run computation-intensive applications, while speeding up computing and saving energy [47]. Computation offloading is to offload the computing tasks of mobile devices to edge servers for computing, so as to solve the shortcomings of mobile devices in terms of storage resources, computing power and energy battery.

Computation offloading has three important decisions to make, namely the offloading decision, the size of the offloaded tasks, and the determination of which tasks to be offloaded [34]. According to the offloading of tasks, computation offloading can be divided into the following three aspects [48, 49]:

- **Completely local execution** : Without offloading to edge servers, all computing tasks are handled by mobile devices.
- **Completely offloading** : All computing tasks are offloaded to edge servers for execution.
- **Partial offloading** : Some computing tasks are processed by mobile devices, and the remaining computing tasks are offloaded to edge servers for processing.

Through investigation and research, we find that the research of computation offloading problem is mainly carried out from three different optimization objectives. The three optimization goals are to minimize latency, minimize energy consumption, and achieve the trade-off between latency and energy consumption.

The offloading decision needs to consider the delay metric, because many applications are delay-sensitive. When the delay is too large, it will seriously affect the user experience of mobile users, and the application may not work properly because the calculation results cannot be obtained in reasonable time. Therefore, the offloading decision needs to guarantee the time delay acceptable to the mobile device. In addition, energy consumption needs to be considered when making offloading decisions. Excessive energy consumption can cause the mobile device's battery to drain rapidly. Minimizing energy consumption is to minimize energy consumption under the premise of satisfying the delay condition. For some applications, latency and energy consumption are both important. At this time, the delay and energy consumption can be weighted and optimized according to the specific needs of the application, so that the total cost is minimized, and the mobile device can obtain the maximum benefit.

1.1.3.1 Minimize Latency

Mobile devices offload computing tasks to edge servers for processing, thereby reducing task processing delays. However, when the amount of tasks to be processed by the mobile device is too large, the delay will inevitably increase. Therefore, it is necessary to find a suitable computation offloading strategy to control the delay.

Liu et al. [35] transformed the computation offloading problem into a Markov decision process and proposed a dynamic offloading algorithm based on Lyapunov optimization. The algorithm could dynamically schedule computing tasks according to the state of the processing unit of the mobile device, the state of the data transmission unit and the queue state of the task buffer. Liu et al.'s approach was to offload tasks to edge servers to reduce execution time on mobile devices. However, the mobile device needs to make the next offloading decision based on the information returned by the edge server. But if the channel state is not good, the mobile device may need to wait for a long time, which inevitably increases the delay time.

Chen et al. [50] studied the mobile edge computation offloading problem in ultra dense networks. Using the idea of software-defined networks, they formulated the task offloading problem as a mixed-integer nonlinear computation process.

Then, they transformed this optimization problem into two sub-problems of task offloading placement problem and resource allocation. Experimental results showed that compared with random offloading tasks and unified offloading tasks, the method could shorten the delay by 20%. Considering the application scenario of the IoT, Yu et al. [51] reduced the computation delay by reasonably allocating resources to the program, and then proposed a completely polynomial time approximation scheme, which could shorten the computation delay more effectively than heuristic algorithms.

Mao et al. [52] proposed a computation offloading strategy for MEC systems based on green energy harvesting [53]. The author took task execution time and task execution failure cost as optimization goals, and proposed a dynamic online offloading algorithm based on Lyapunov optimization. The algorithm only needed to consider the current system state when making the offloading decision. Real-time allocation of system resources was realized, which greatly reduced the execution delay of the system.

In the cloud-edge collaborative task offloading system, the offloading of computing tasks from the network edge to the remote transport center usually leads to high latency, which is very unfriendly to some delay-sensitive applications [54–56]. Therefore, it is very important to design an appropriate computation offloading strategy to optimize the delay of task offloading. In order to solve the problem of limited computing power of MEC servers, Wang et al. [57] proposed a heterogeneous multi-layer MEC system. The idea of their design of this system was: if the task generated by the mobile device couldn't be executed locally in time, the task was offloaded to the edge server for execution; if the edge server was still unable to process the data offloaded from the mobile device in time, the task could be offloaded to a cloud center with stronger processing power for processing. They minimized system latency by jointly handling task distribution among edge devices, multi-tier MEC servers, and cloud computing centers.

The above computation offloading decisions all achieve the purpose of shortening the delay during the offloading process, but the energy consumption on the side of the mobile terminal device is not considered when making the offloading decision. Therefore, scholars continue to explore computation offloading decisions that can minimize energy consumption while meeting the delay constraints of mobile terminal devices.

1.1.3.2 Minimize Energy Consumption

In Sect. 1.1.3.1, we present the computation offloading schemes with the optimization goal of minimizing the delay. However, there are also some application scenarios that are more concerned with energy consumption within an acceptable delay range. Next, we introduce the computation offloading scheme with the optimization goal of minimizing energy consumption.

You et al. [58] considered offloading computing tasks generated by multiple mobile devices to a single edge server for processing. To this end, they proposed a resource allocation problem to minimize system energy consumption under time delay constraints. They prioritized mobile devices for task offloading based on their channel gain and local computation energy consumption. At the same time, they also designed a threshold for the energy consumption generated by performing the task. When the priority of the mobile device was higher than the threshold, a complete offloading strategy is implemented. When the priority of the mobile device was lower than the threshold, the partial offloading strategy was implemented. By using this strategy, there was see a significant improvement in system performance. Due to the limited resources of mobile devices, Dbouk et al. [59] proposed a multi-objective resource-aware optimization model. Then, they used genetic algorithms to achieve dynamic analysis of mobile device statistics and came up with smart offloading schemes.

Different from the above schemes, Lyu et al. [60] proposed a new system architecture, that was, a system integrating remote cloud, edge servers and IoT devices. They proposed a framework that allowed lightweight requests to solve the proposed scalability problem. Since there was no coordination between devices, the framework could run separately on mobile devices and servers by encapsulating latency requirements in offload requests. Then, a selective offloading scheme was designed to minimize the energy consumption of the equipment. Signaling overhead could be further reduced by allowing devices to self-designate or self-reject offload. The advantage of this solution was that the device could choose whether to offload it according to its actual situation.

Zhao et al. [61] conducted a computation offloading study of a multi-mobile user MEC system. To minimize the energy consumption of mobile devices, they not only considered offloading options, but also jointly optimized the allocation of radio and computing resources. They formulated the energy minimization problem as a mixed integer nonlinear programming problem and proposed a heuristic greedy algorithm based on Gini coefficients to solve this problem.

Cao et al. [62] conducted a computation offloading study of a three-node MEC system. Under time delay constraints, they proposed a protocol for joint computation and communication allocation. They jointly optimized the task division and time allocation of terminal nodes and auxiliary nodes to minimize the total energy consumption under the time delay constraint. But they only considered the classic three-node MEC system. In practical applications, the number of nodes is much larger than three.

Zhao et al. [63] studied task offloading for heterogeneous cloud-based multi-user MEC systems. The mobile device transmitted the generated tasks to the edge cloud through a wireless channel, and the tasks arriving at the edge cloud could be further transmitted to the remote cloud through the Internet. They proposed a joint assignment algorithm based on dynamic programming. The algorithm could synchronously allocate bandwidth and computing resources, aiming to minimize the total energy consumption of multiple mobile devices.

Computation offloading decisions that minimize energy consumption generally seek to minimize energy consumption while satisfying acceptable latency constraints for mobile devices. Such a strategy of minimizing energy consumption depends on the transmit power of the mobile device and the quality of the wireless channel. Most of the existing strategies are verified by means of simulation, and conclusions are obtained. However, it is difficult to restore the real offloading conditions in the simulation experiment, and the time-varying wireless channel quality and the mutual interference when multiple mobile devices offload at the same time may be ignored. In a scenario where multiple mobile devices use the same MEC server for computation offloading, it is generally assumed that all mobile devices have the same channel quality and computing capability, which is far from the real situation. In the actual offloading scenarios, it may be necessary to minimize both the delay and the energy consumption. But it is unrealistic to minimize the two conflicting metrics at the same time. Therefore, researchers minimize the weighted sum of the two metrics. By balancing latency and energy consumption, the maximum benefit of computation offloading is achieved.

1.1.3.3 Weighted Sum of Latency and Energy Consumption

In the above, we discuss the offloading schemes with the optimization goals of minimizing the delay and minimizing energy consumption, respectively. However, in some scenarios, it may be necessary to optimize both latency and energy consumption, and the optimal goal will be to optimize the weighted sum of latency and energy consumption.

Chen et al. [64] studied the multi-user computation offloading problem in a multi-channel wireless interference environment. The author proved that the computation offloading problem to be solved was NP-hard, and then designed the offloading strategy according to the characteristics of distributed systems and game theory. The author proved that the proposed model had a Nash equilibrium, and designed a distributed computation offloading algorithm. The computation offloading scheme used classical game theory to solve the problem, and the algorithm had high fitness and could maintain good performance in different scenarios.

Dinh et al. [65] studied the problem of computation offloading under multi-edge servers. They considered offloading tasks generated by a single mobile device to multiple edge servers for processing. For the case of the CPU cycle frequency of mobile devices, they considered fixed and non-fixed, respectively. Then, for the two cases, they proposed corresponding computation offloading methods respectively. The comparison results showed that the CPU frequency of mobile devices had a certain impact on task allocation.

Considering the heterogeneity between the edge server and the remote cloud, Wu et al. [66] established a hybrid offloading model combining MCC and MEC. They proposed an offloading algorithm for distributed computation based on deep learning. The algorithm could combine mobile devices, edge servers and cloud servers to generate near-optimal offloading decisions.

There is a big gap in resource capacity between MEC and cloud. Therefore, in order to achieve a balance between performance and cost, Zhou et al. [67] proposed an online cloud-side resource provisioning framework based on latency-aware

Lyapunov optimization techniques. The framework could allocate MEC resources and cloud resources to edge users with optimal strategies to achieve a trade-off between latency and cost.

The computation offloading strategy that maximizes revenue is essentially to analyze the influence of the two indicators of delay and energy consumption on the total energy consumption of computation offloading in the process of performing computation offloading under the condition that the execution time limit and energy consumption are satisfied, so as to reduce the cost of computation offloading. The two are weighed and a balance point is found to make the limitation of delay or energy consumption more suitable for the actual scene, so as to maximize the profit.

1.2 Challenges

MEC can significantly reduce the energy consumption of mobile terminal devices, and can meet the requirements of real-time application offloading, so it has received extensive attention. Computation offloading technology can help mobile devices to better process computing tasks. However, since the computation offloading technology is still immature, there are still many problems to be solved before applying it to the mobile network. In this section, we discuss the challenges faced by computation offloading. We illustrate the challenges in computation offloading from the following six aspects.

(A) Stochastic Task Generation and Arrival Process

One of the challenges of computation offloading is that task generation and arrival processes are stochastic and hard to predict precisely. IoT devices are influenced by user needs when generating tasks. The device may not generate new tasks when there is no demand from the user. When users have frequent service requests, the device may generate a large number of tasks. Therefore, the generation of tasks for mobile devices is stochastic. We also cannot predict the arrival process of tasks. Mobile users may request services at any time. That is, mobile devices may generate new tasks at any time. Therefore, the arrival process of the task is stochastic, and we cannot accurately predict the arrival process of the task. Not only that, it is impractical to estimate accurately the size of the generated tasks.

(B) Dynamics and Uncertainty of Wireless Channel States

The design of the computation offloading strategy is also affected by the wireless channel state.

First, the wireless channel state is dynamic. The time-varying wireless channel state greatly affects the optimal offloading decision of the system. When the wireless channel state is not good, the offloading of tasks will take more time and bring greater energy consumption. When the wireless channel is in good condition, the offloading of tasks will become faster, which will save energy. However, the state of the wireless channel is dynamic, that is, the state of the wireless channel is constantly changing all the time.

Second, the state of the wireless channel is also uncertain. On the one hand, the state of the wireless channel is not only affected by the location of IoT devices, but also by network congestion and fading. On the other hand, the state of the wireless channel will change with the changes of internal factors of the MEC system and many external environments. The external environment includes weather, temperature, etc. For example, if the weather is bad, it may cause the channel quality to deteriorate.

(C) Tradeoff Between Energy Consumption and Latency
When designing offloading decisions, how to balance energy consumption and latency is also very important. Sometimes, for applications that are not delay-sensitive, reducing energy consumption as much as possible will be the first choice. However, for delay-sensitive applications such as real-time communications, how to optimize energy consumption may not be so important. With the development of computation-intensive applications, people begin to demand not only low latency, but also low energy consumption. But it is almost impossible to meet both requirements at the same time. Therefore, only a compromise can be taken to achieve a tradeoff between energy consumption and latency.

D. Underutilization of Resources
Improving the utilization rate of resources is also one of the important challenges in computation offloading for MEC. Mobile devices are selfish, and each device wants to minimize the delay of its task or minimize energy consumed by processing tasks. Therefore, the competition of resources will be performed between mobile devices. At the same time, due to the uneven spatial distribution of mobile devices and edge servers, the workload between edge servers is unbalanced. Therefore, how to ensure that mobile device offload the tasks to the appropriate edge servers and how to balance the workload between the edge servers, is a challenge that needs to be addressed. When allocating resources, it is desirable that the mobile device with a large amount of offloading tasks can obtain more resources to meet the computation demand of the mobile device.

E. Heterogeneity of Devices and Servers
With the development of 5G technology, IoT devices are also developing rapidly, accompanied by the emergence of various emerging applications. However, due to the complex and diverse mobile network environment, the heterogeneity of servers and devices is also an important issue.

First, IoT devices are heterogeneous. Various IoT devices are different in structure, and devices produced by different manufacturers are almost all different. At the same time, different IoT devices access the network in different ways.

Second, edge servers are heterogeneous. Today, there are many types of edge servers. Each edge server is designed to meet one or several services. In practical applications, in order to better meet the needs of users, multiple servers are required to be used together. Therefore, it is also very important to solve the heterogeneity between edge servers. Some existing computation offloading schemes assume the same structure of MEC servers, which is inconsistent with the heterogeneity of the actual network.

Third, the difference in processing power between IoT devices, edge servers and cloud servers is huge. Generally, the processing power of cloud servers is greater than that of edge servers, and the processing power of IoT devices is the smallest. When designing an offloading strategy, the difference in processing power among the three should be considered at the same time. Therefore, when offloading tasks, the difference in computing power between IoT devices and edge servers and cloud servers should also be considered.

Therefore, how to design computation offloading strategy considering the heterogeneities between different types of devices, between different servers, and between devices and servers is a severe challenge.

(F) Large-Scale Search Space and State Space

With the development of mobile computing technology, more and more new applications are emerging, and the number of mobile devices is also growing rapidly. In response to the proliferation of computing tasks brought by the promotion of mobile devices, the number of servers has also increased. As the number of IoT devices and servers continues to grow at the same time, this also leads to an increase in the search space and state space. This directly leads to the challenges of accuracy and solution efficiency.

In general, two methods are used for computation offloading solutions, the centralized method and the heuristic method. The centralized optimization method can achieve precise computation offloading. However, with the increase in the number of IoT devices and base stations, the search space and state space also increase, which will lead to lower efficiency of finding the solution. In contrast, heuristic algorithms can achieve efficient computation offloading. However, since the heuristic algorithm is tried in a limited search space, its accuracy and precision cannot meet the established requirements. Therefore, we need an optimization method that can achieve a compromise between accuracy, precision, and efficiency.

1.3 Contributions

The contributions of this book are organized as follows.

Firstly, we provide a comprehensive review on the background of MCC and MEC. The characteristics and typical applications of MCC and MEC are also presented. Furthermore, the background and motivations of computation offloading are discussed. The challenges in computation offloading in MEC are summerized.

Secondly, the transmission energy consumption optimization problem is studied in computation offloading in MEC. Specifically, with the rapid development of the IoT, more and more computation-intensive applications appeare and generate a large number of computation-intensive tasks. However, due to the limitations of mobile devices themselves, they cannot fully realize local computation. Therefore, the task needs to be offloaded to the MEC server at the network edge for computing. However, offloading these tasks to edge servers for processing consumes a lot of energy. Therefore, we formulate the task offloading optimization problem with the goal of minimizing the transmission energy consumption. However, since we cannot accurately predict the statistical information of task generation process of mobile devices and the state of the wireless channel, we employ stochastic optimization techniques to solve this problem. We design an energy efficient dynamic computation offloading (EEDCO) algorithm based on stochastic optimization techniques. Our EEDCO algorithm does not require prior statistics of task generation and arrival process. The EEDCO algorithm can be implemented in a parallel way with low complexity. Experiments show that our EEDCO algorithm can minimize the transmission energy consumption while reducing the queue length.

Thirdly, the energy efficient offloading and frequency scaling problem for IoT devices in MEC is addressed. For the IoT devices in the MEC system, one of the main challenges is the limited battery capacity, which cannot support the processing of computation-intensive tasks for a long time. When a mobile device processes computing tasks locally, the amount of energy consumed is mainly related to the local CPU cycle frequency. When a mobile device performs task offloading, the energy consumed is mainly related to the amount of offloaded tasks and the transmission power of the device. In addition, the device still runs for a period of time (tail time) after completing the data transfer. Therefore, we also consider tail energy consumption. Similarly, the generation and arrival process of tasks are random and the statistical information is hard to predict precisely. The wireless channel state is also unpredictable. Therefore, we adopt the stochastic optimization techniques and transform the original joint task assignment and frequency scaling problem into a deterministic optimization problem. Then, a Computation Offloading and Frequency Scaling for Energy Efficiency (COFSEE) algorithm is proposed for the IoT devices in the MEC system.

Fourthly, Deep Reinforcement Learning-based energy efficient computation offloading solutions for MEC systems with multiple edge servers are proposed. Mobile devices dynamically generate computing tasks and offload them to edge servers for processing through wireless channels. However, mobile devices tend to choose edge servers that can provide themselves with the best computing resources. At the same time, due to the delay-sensitive nature of mobile devices, in order to complete more tasks under delay constraints and minimize system energy consumption, we propose an end-to-end deep reinforcement learning algorithm. The algorithm jointly optimizes the allocation of computing resources and the selection of edge servers. Our algorithm maximizes long-term cumulative returns. Experiments demonstrate that our algorithm can achieve energy efficient computation offloading and maximize system utility.

Fifthly, the solutions of energy efficient computation offloading in NOMA-based MEC are proposed. We consider two cases. The former case is that the channel state is static. For this case, a hierarchical algorithm is proposed that jointly optimizes NOMA transmission, resource allocation, and multi-access multitasking computation offloading. For the latter case, we consider the channel power gain from the mobile device to the edge server is time-varying. To solve the problems brought by different channel implementations in dynamic scenarios, we propose an online algorithm based on DRL. The algorithm can achieve an approximate optimal unloading solution through continuous learning. Experimental results demonstrate the effectiveness of the proposed algorithm.

1.4 Book Outline

The outline of the book is as follows.

In the first chapter, we present the background of MCC and MEC. The motivation of computation offloading is also given. Then, the problems encountered in computation offload are discussed.

In the second chapter, we study the transmission energy optimization for task offloading in MEC. Since the processes of task generation and arrival and the state of the wireless channels are dynamic, we apply stochastic optimization techniques to propose the EEDCO algorithm for solving the problem. The EEDCO algorithm can achieve the flexible tradeoff between energy consumption and queue length, and minimize the energy consumption while bounding the queue length.

In the third chapter, we study energy efficient offloading and CPU frequency scaling problems for IoT devices. To improve the energy efficiency of computation offloading with tail energy consumption, we jointly optimize computing task allocation and CPU cycle frequency scaling, and propose the COFSEE algorithm. COFSEE can achieve minimal energy consumption while ensuring bounded queue length.

In the fourth chapter, we consider delay-aware and energy efficient computation offloading in a dynamic MEC system with multiple edge servers. In order to select the best edge server for offloading and allocate the optimal computing resource such that the expected long-term utility is maximized, we propose a scheme of end-to-end DRL.

In the fifth chapter, we study the problem of energy-efficient computation offloading with NOMA scenarios. We propose a hierarchical algorithm for the static case and DRL algorithm for the dynamic case.

In the sixth chapter, we conclude the book and give directions for future research.

References

1. A. Gupta, R.K. Jha, A survey of 5g network: architecture and emerging technologies. IEEE Access **3**, 1206–1232 (2015). https://doi.org/10.1109/ACCESS.2015.2461602
2. G.A. Akpakwu, B.J. Silva, G.P. Hancke, A.M. Abu-Mahfouz, A survey on 5G networks for the internet of things: communication technologies and challenges. IEEE Access **6**, 3619–3647 (2018). https://doi.org/10.1109/ACCESS.2017.2779844

3. Y. Siriwardhana, P. Porambage, M. Liyanage, M. Ylianttila, A survey on mobile augmented reality with 5G mobile edge computing: architectures, applications, and technical aspects. IEEE Commun. Surv. Tuts. **23**(2), 1160–1192 (2021). https://doi.org/10.1109/COMST.2021.3061981
4. A.u.R. Khan, M. Othman, S.A. Madani, S.U. Khan, A survey of mobile cloud computing application models. IEEE Commun. Surv. Tuts. **16**(1), 393–413 (2014). https://doi.org/10.1109/SURV.2013.062613.00160
5. S. Wang, J. Xu, N. Zhang, Y. Liu, A survey on service migration in mobile edge computing. IEEE Access **6**, 23511–23528 (2018). https://doi.org/10.1109/ACCESS.2018.2828102
6. Y. Mao, C. You, J. Zhang, K. Huang, K.B. Letaief, A survey on mobile edge computing: the communication perspective. IEEE Commun. Surv. Tuts. **19**(4), 2322–2358 (2017). https://doi.org/10.1109/COMST.2017.2745201
7. H. Li, G. Shou, Y. Hu, Z. Guo, Mobile edge computing: progress and challenges, in *2016 4th IEEE International Conference on Mobile Cloud Computing, Services, and Engineering (MobileCloud)* (2016), pp. 83–84. https://doi.org/10.1109/MobileCloud.2016.16
8. M. Armbrust et al., A view of cloud computing. Commun. ACM **53**(4), 50–58 (2010)
9. I. Sahu, U.S. Pandey, Mobile cloud computing: issues and challenges, in *2018 International Conference on Advances in Computing, Communication Control and Networking (ICACCCN)* (2018), pp. 247–250. https://doi.org/10.1109/ICACCCN.2018.8748376
10. R.-S. Chang, J. Gao, V. Gruhn, J. He, G. Roussos, W.-T. Tsai, Mobile cloud computing research - issues, challenges and needs, in *2013 IEEE Seventh International Symposium on Service-Oriented System Engineering*, (2013), pp. 442–453. https://doi.org/10.1109/SOSE.2013.96
11. G. Kumar, E. Jain, S. Goel, V.K. Panchal, Mobile cloud computing architecture, application model, and challenging issues, in *2014 International Conference on Computational Intelligence and Communication Networks* (2014), pp. 613–617. https://doi.org/10.1109/CICN.2014.137
12. Z. Pang, L. Sun, Z. Wang, E. Tian, S. Yang, A survey of cloudlet based mobile computing, in *2015 International Conference on Cloud Computing and Big Data (CCBD)* (2015), pp. 268–275. https://doi.org/10.1109/CCBD.2015.54
13. Q. Fan, L. Liu, A survey of challenging issues and approaches in mobile cloud computing, in *2016 17th International Conference on Parallel and Distributed Computing, Applications and Technologies (PDCAT)*, (2016), pp. 87–90. https://doi.org/10.1109/PDCAT.2016.032
14. S. Goyal, J. Carter, A lightweight secure cyber foraging infrastructure for resource-constrained devices, in *Sixth IEEE Workshop on Mobile Computing Systems and Applications* (2004), pp. 186–195. https://doi.org/10.1109/MCSA.2004.2
15. M. Satyanarayanan, P. Bahl, R. Caceres, N. Davies, The case for VM-based cloudlets in mobile computing. IEEE Pervasive Comput. **8**(4), 14–23 (2009). https://doi.org/10.1109/MPRV.2009.82
16. Y. Jararwah, L. Tawalbeh, F. Ababneh, A. Khreishah, F. Dosari, Scalable cloudlet-based mobile computing model, in *The 11th International Conference on Mobile Systems and Pervasive Computing-Mobi-SPC 2014, Niagara Falls, August 17–20*. Procedia Computer Science, vol. 34 (Elsevier, Amsterdam, 2014), pp. 434–441
17. Y. Li, W. Wang, Can mobile cloudlets support mobile applications? in *IEEE INFOCOM 2014 - IEEE Conference on Computer Communications* (2014), pp. 1060–1068. https://doi.org/10.1109/INFOCOM.2014.6848036
18. T. Soyata, R. Muraleedharan, C. Funai, M. Kwon, W. Heinzelman, Cloud-vision: real-time face recognition using a mobile-cloudlet-cloud acceleration architecture, in *2012 IEEE Symposium on Computers and Communications (ISCC)* (2012), pp. 000059–000066. https://doi.org/10.1109/ISCC.2012.6249269
19. L.A. Tawalbeh, W. Bakhader, R. Mehmood, H. Song, Cloudlet-based mobile cloud computing for healthcare applications, in *2016 IEEE Global Communications Conference (GLOBECOM)* (2016), pp. 1–6. https://doi.org/10.1109/GLOCOM.2016.7841665

20. F. Bonomi, R. Milito, J. Zhu, S. Addepalli, Fog computing and its role in the internet of things, in *Proceedings of the First Edition of the MCC Workshop on Mobile Cloud Computing* (ACM, New York, 2012), pp. 13–16
21. M.R. Raza, A. Varol, N. Varol, Cloud and fog computing: a survey to the concept and challenges, in *2020 8th International Symposium on Digital Forensics and Security (ISDFS)* (2020), pp. 1–6. https://doi.org/10.1109/ISDFS49300.2020.9116360
22. R. Mahmud, R. Kotagiri, R. Buyya, Fog computing: a taxonomy, survey and future directions. arXiv:1611.05539v4 [cs.DC] (2017)
23. M. Mukherjee, L. Shu, D. Wang, Survey of fog computing: fundamental, network applications, and research challenges. IEEE Commun. Surv. Tuts. **20**(3), 1826–1857 (2018). https://doi.org/10.1109/COMST.2018.2814571
24. P. Zhang, M. Zhou, G. Fortino, Security and trust issues in fog computing: a survey. Future Gener. Comput. Syst. **88**, 16–27 (2018)
25. M. La Polla, F. Martinelli, D. Sgandurra, A survey on security for mobile devices. IEEE Commun. Surv. Tuts. **15**(1), 446–471 (2013). https://doi.org/10.1109/SURV.2012.013012.00028
26. B. Wang, B. Li, H. Li, Oruta: privacy-preserving public auditing for shared data in the cloud. IEEE Trans. Cloud Comput. **2**(1), 43–56 (2014). https://doi.org/10.1109/TCC.2014.2299807
27. A. Shabtai, U. Kanonov, Y. Elovici, Intrusion detection for mobile devices using the knowledge-based, temporal abstraction method. J. Syst. Softw. **83**, 1524–1537 (2010)
28. I. Khalil, A. Khreishah, M. Azeem, Consolidated identity management system for secure mobile cloud computing. Comput. Netw. **65**, 99–110 (2014)
29. P. Gupta, S. Gupta, Mobile cloud computing: the future of cloud[J]. Int. J. Adv. Res. Electr. Electron. Instrum. Eng. **1**(3), 134–145 (2012)
30. N. Abbas, Y. Zhang, A. Taherkordi, T. Skeie, Mobile edge computing: a survey. IEEE Internet Things J. **5**(1), 450–465 (2018). https://doi.org/10.1109/JIOT.2017.2750180
31. J. Zhao, Y. Liu, Y. Gong, C. Wang, L. Fan, A dual-link soft handover scheme for C/U plane split network in high-speed railway. IEEE Access **6**, 12473–12482 (2018). https://doi.org/10.1109/ACCESS.2018.2794770
32. D. Chatzopoulos, C. Bermejo, S. Kosta, P. Hui, Offloading computations to mobile devices and cloudlets via an upgraded NFC communication protocol. IEEE Trans. Mobile Comput. **19**(3), 640–653 (2020). https://doi.org/10.1109/TMC.2019.2899093
33. W. Li, Z. Chen, X. Gao, W. Liu, J. Wang, Multimodel framework for indoor localization under mobile edge computing environment. IEEE Internet Things J. **6**(3), 4844–4853 (2019). https://doi.org/10.1109/JIOT.2018.2872133
34. Y. Zhang, H. Liu, L. Jiao, X. Fu, To offload or not to offload: an efficient code partition algorithm for mobile cloud computing, in *2012 IEEE 1st International Conference on Cloud Networking (CLOUDNET)* (2012), pp. 80–86. https://doi.org/10.1109/CloudNet.2012.6483660
35. J. Liu, Y. Mao, J. Zhang, K.B. Letaief, Delay-optimal computation task scheduling for mobile-edge computing systems, in *2016 IEEE International Symposium on Information Theory (ISIT)* (2016), pp. 1451–1455. https://doi.org/10.1109/ISIT.2016.7541539
36. T.H. Luan, L. Gao, Z. Li, Y . Xiang, L. Sun, Fog computing: focusing on mobile users at the edge. CoRR, abs/1502.01815 (2015)
37. ETSI M., Mobile edge computing (mec); framework and reference architecture[J]. ETSI, DGS MEC (2016)
38. T. Wang, G. Zhang, A. Liu, M.Z.A. Bhuiyan, Q. Jin, A secure IoT service architecture with an efficient balance dynamics based on cloud and edge computing. IEEE Internet Things J. **6**(3), 4831–4843 (2019). https://doi.org/10.1109/JIOT.2018.2870288
39. Y. Mao, C. You, J. Zhang, K. Huang, K.B. Letaief, A survey on mobile edge computing: the communication perspective. IEEE Commun. Surv. Tuts. **19**(4), 2322–2358 (2017). https://doi.org/10.1109/COMST.2017.2745201

40. Y. Chen, Y. Zhang, Y. Wu, L. Qi, X. Chen, X. Shen, Joint task scheduling and energy management for heterogeneous mobile edge computing with hybrid energy supply. IEEE Internet Things J. **7**(9), 8419–8429 (2020). https://doi.org/10.1109/JIOT.2020.2992522
41. Y. Deng, Z. Chen, X. Yao, S. Hassan, J. Wu, Task scheduling for smart city applications based on multi-server mobile edge computing. IEEE Access **7**, 14410–14421 (2019). https://doi.org/10.1109/ACCESS.2019.2893486
42. X. Xu, Q. Huang, X. Yin, M. Abbasi, M.R. Khosravi, L. Qi, Intelligent offloading for collaborative smart city services in edge computing. IEEE Internet Things J. **7**(9), 7919–7927 (2020). https://doi.org/10.1109/JIOT.2020.3000871
43. C. Liu et al., A new deep learning-based food recognition system for dietary assessment on an edge computing service infrastructure. IEEE Trans. Serv. Comput. **11**(2), 249–261 (2018). https://doi.org/10.1109/TSC.2017.2662008
44. N. Kumar, S. Zeadally, J.J.P.C. Rodrigues, Vehicular delay-tolerant networks for smart grid data management using mobile edge computing. IEEE Commun. Mag. **54**(10), 60–66 (2016). https://doi.org/10.1109/MCOM.2016.7588230
45. Y. Siriwardhana, P. Porambage, M. Liyanage, M. Ylianttila, A survey on mobile augmented reality with 5G mobile edge computing: architectures, applications, and technical aspects. IEEE Commun. Surv. Tuts. **23**(2), 1160–1192 (2021). https://doi.org/10.1109/COMST.2021.3061981
46. C. Jiang, X. Cheng, H. Gao, X. Zhou, J. Wan, Toward computation offloading in edge computing: a survey. IEEE Access **7**, 131543–131558 (2019). https://doi.org/10.1109/ACCESS.2019.2938660
47. S. Kumar, M. Tyagi, A. Khanna, V. Fore, A survey of mobile computation offloading: applications, approaches and challenges, in *2018 International Conference on Advances in Computing and Communication Engineering (ICACCE)* (2018), pp. 51–58. https://doi.org/10.1109/ICACCE.2018.8441740
48. P. Mach, Z. Becvar, Mobile edge computing: a survey on architecture and computation offloading. IEEE Commun. Surv. Tuts. **19**(3), 1628–1656 (2017). https://doi.org/10.1109/COMST.2017.2682318
49. P. Mach, Z. Becvar, Mobile edge computing: a survey on architecture and computation offloading. IEEE Commun. Surv. Tuts. **19**(3), 1628–1656 (2017). https://doi.org/10.1109/COMST.2017.2682318
50. M. Chen, Y. Hao, Task offloading for mobile edge computing in software defined ultra-dense network. IEEE J. Sel. Areas Commun. **36**(3), 587–597 (2018). https://doi.org/10.1109/JSAC.2018.2815360
51. R. Yu, G. Xue, X. Zhang, Application provisioning in FOG computing-enabled internet-of-things: a network perspective, in *IEEE INFOCOM 2018 - IEEE Conference on Computer Communications* (2018), pp. 783–791. https://doi.org/10.1109/INFOCOM.2018.8486269
52. Y. Mao, J. Zhang, K.B. Letaief, Dynamic computation offloading for mobile-edge computing with energy harvesting devices. IEEE J. Sel. Areas Commun. **34**(12), 3590–3605 (2016). https://doi.org/10.1109/JSAC.2016.2611964
53. S. Ulukus, et al., Energy harvesting wireless communications: a review of recent advances. IEEE J. Sel. Areas Commun. **33**(3), 360–381 (2015). https://doi.org/10.1109/JSAC.2015.2391531
54. W. Zhang, S. Li, L. Liu, Z. Jia, Y. Zhang, D. Raychaudhuri, Hetero-edge: orchestration of real-time vision applications on heterogeneous edge clouds, in *IEEE INFOCOM 2019 - IEEE Conference on Computer Communications* (2019), pp. 1270–1278. https://doi.org/10.1109/INFOCOM.2019.8737478
55. L.I. Carvalho, D.M.A. da Silva, R.C. Sofia, Leveraging context-awareness to better support the IoT cloud-edge continuum, in *2020 Fifth International Conference on Fog and Mobile Edge Computing (FMEC)* (2020), pp. 356–359. https://doi.org/10.1109/FMEC49853.2020.9144760
56. P. Wang, Z. Zheng, B. Di, L. Song, Joint task assignment and resource allocation in the heterogeneous multi-layer mobile edge computing networks, in *2019 IEEE Global Communications*

Conference (GLOBECOM) (2019), pp. 1–6. https://doi.org/10.1109/GLOBECOM38437.2019.9014309
57. P. Wang, Z. Zheng, B. Di, L. Song, HetMEC: latency-optimal task assignment and resource allocation for heterogeneous multi-layer mobile edge computing. IEEE Trans. Wireless Commun. **18**(10), 4942–4956 (2019). https://doi.org/10.1109/TWC.2019.2931315
58. C. You, K. Huang, Multiuser resource allocation for mobile-edge computation offloading, in *2016 IEEE Global Communications Conference (GLOBECOM)* (2016), pp. 1–6. https://doi.org/10.1109/GLOCOM.2016.7842016
59. T. Dbouk, A. Mourad, H. Otrok, H. Tout, C. Talhi, A novel ad-hoc mobile edge cloud offering security services through intelligent resource-aware offloading, IEEE Trans. Netw. Serv. Manag. **16**(4), 1665–1680 (2019). https://doi.org/10.1109/TNSM.2019.2939221
60. X. Lyu et al., Selective offloading in mobile edge computing for the green Internet of Things. IEEE Netw. **32**(1), 54–60 (2018). https://doi.org/10.1109/MNET.2018.1700101
61. P. Zhao, H. Tian, C. Qin, G. Nie, Energy-saving offloading by jointly allocating radio and computational resources for mobile edge computing. IEEE Access **5**, 11255–11268 (2017). https://doi.org/10.1109/ACCESS.2017.2710056
62. X. Cao, F. Wang, J. Xu, R. Zhang, S. Cui, Joint computation and communication cooperation for mobile edge computing, in *2018 16th International Symposium on Modeling and Optimization in Mobile, Ad Hoc, and Wireless Networks (WiOpt)* (2018), pp. 1–6. https://doi.org/10.23919/WIOPT.2018.8362865
63. T. Zhao, S. Zhou, L. Song, Z. Jiang, X. Guo, Z. Niu, Energy-optimal and delay-bounded computation offloading in mobile edge computing with heterogeneous clouds. China Commun. **17**(5), 191–210 (2020). https://doi.org/10.23919/JCC.2020.05.015
64. X. Chen, L. Jiao, W. Li, X. Fu, Efficient multi-user computation offloading for mobile-edge cloud computing. IEEE/ACM Trans. Netw. **24**(5), 2795–2808 (2016). https://doi.org/10.1109/TNET.2015.2487344
65. T.Q. Dinh, J. Tang, Q.D. La, T.Q.S. Quek, Offloading in mobile edge computing: task allocation and computational frequency scaling. IEEE Trans. Commun. **65**(8), 3571–3584 (2017). https://doi.org/10.1109/TCOMM.2017.2699660
66. H. Wu, Z. Zhang, C. Guan, K. Wolter, M. Xu, Collaborate edge and cloud computing with distributed deep learning for smart city Internet of Things. IEEE Internet Things J. **7**(9), 8099–8110 (2020). https://doi.org/10.1109/JIOT.2020.2996784
67. Z. Zhou, S. Yu, W. Chen, X. Chen, CE-IoT: cost-effective cloud-edge resource provisioning for heterogeneous IoT applications. IEEE Internet Things J. **7**(9), 8600–8614 (2020). https://doi.org/10.1109/JIOT.2020.2994308

Chapter 2
Dynamic Computation Offloading for Energy Efficiency in Mobile Edge Computing

2.1 System Model and Problem Statement

In recent years, with the rapid development of the IoT and 5G techniques, the types of IoT services have become more and more diverse, and the number of computation-intensive applications running on the IoT devices has also been increasing dramatically [1, 2]. These computation-intensive applications would consume a lot of energy. Due to the limited computing power and battery capacity of IoT devices, the IoT devices are not powerful enough to process all the data generated, and processing them on local IoT devices will degrade the performance. In order to solve this problem, the computing tasks of IoT devices can be offloaded to the cloud with powerful computing processing capabilities. Besides, MEC is an emerging technology that can use wireless access networks to provide users with required services and cloud computing functions nearby. The distance between traditional cloud computing and IoT devices is generally very long, but edge servers can be deployed on wireless access points such as base stations near the IoT devices. MEC can reduce core network traffic and service delay [3]. By offloading computing tasks [4], IoT devices can obtain better computing services and extend battery life. Therefore, the task offloading of from IoT devices to MEC has attracted great attention from the industry and academia [5, 6].

Offloading computing tasks from IoT devices to MEC still generate high energy consumption of the devices, which accounts for a large part of the total energy consumption of the devices [7, 8]. For IoT devices, the transmission energy consumption of each device is greatly affected by the state of the wireless channel. The transmission rate of IoT device is closely related to the quality of the channel conditions. The transmission rate of IoT device is higher when the channel conditions are better, thereby reducing data transmission time and reducing transmission energy consumption. When transmitting the same amount of data, when the channel conditions are worse, the energy consumed for transmitting the data will increase. In order to reduce the transmission energy consumption of IoT

Y. Chen et al., *Energy Efficient Computation Offloading in Mobile Edge Computing*, Wireless Networks, https://doi.org/10.1007/978-3-031-16822-2_2

devices, tasks can be offloaded when the channel conditions are good. However, this might lead to a very large backlog of tasks on IoT devices, and make the queue length very large and unstable [9]. Therefore, it is necessary to design an efficient task offloading strategy that minimizes offloading energy consumption while guarantee quality of services (QoS) to IoT devices.

It is difficult to design such an effective strategy to meet the above-mentioned requirements. First, the quality of the wireless channel cannot be accurately or precisely predicted because it changes dynamically with time. Second, the location of the IoT device will also affect its quality. The degree of network congestion and fading will also affect it. At the same time, the statistical information of task arrival process for each IoT device is difficult to obtain. Therefore, it is very challenging to design such an energy efficient offloading strategy that can not only adapt to the dynamic changes of channel conditions and the dynamic arrival process of tasks, but also reduces the energy consumption while guarantee the QoS. Moreover, with the rapid increase in the number of IoT devices, the scale of the energy-efficient dynamic offloading problem will be huge [10]. Therefore, in order to solve this problem, it is necessary to design an efficient and low-complexity offloading algorithm.

2.1.1 Network Model

We consider there is a base station (BS) in the MEC system. The BS is equipped with an MEC server to provide services for n IoT devices nearby. IoT devices can offload computing tasks to the MEC server for processing. The IoT device connects to the MEC server through a wireless channel and offloads tasks. In this way, IoT devices not only can obtain better service quality, but also extend the battery life. IoT devices set is defined by $I = \{1, 2, \cdots, n\}$. A time-slotted system is defined by $t \in \{0, 1, \cdots, T-1\}$, and the length of time slot is τ. The main symbols are given in Table 2.1.

2.1.2 Task Offloading Model

In order to be general, IoT devices are considered to be heterogeneous. And the amount of task data that the IoT device i generates in the time slot t is denoted as $A_i(t)$ (in bits). Therefore, $A_i(t)$ are different for different IoT devices. It is worth noting that the prior statistical information of $A_i(t)$ is not needed here, and it is difficult to obtain in the actual systems.

The available number of uplink sub-channels is expressed as $S(t)$. To be more general, we suppose that $S(t)$ changes dynamically in different time slots. The transmission power of IoT device i is defined as P_i. The channel power gain of IoT device i in time slot t is defined as $h_i(t)$. Then, the achievable task offload rate

Table 2.1 Notations and definitions

Notion	Definition
I	IoT devices set
τ	Time slot length
P_i	Transmit power of the IoT device i
$S(t)$	Number of available sub-channels in slot t
$h_i(t)$	Channel power gain for IoT device i on the sub-channel in time slot t
B	Bandwidth of each sub-channel
N_0	Noise power spectral density
$R_i(t)$	Task offloading rate of IoT device i in slot t
$\kappa_i(t)$	Offloading duration of IoT device i during slot t
$e(t)$	Transmission energy consumption of all IoT devices in slot t
$A_i(t)$	Amount of computation tasks arrived at IoT device i in slot t
$Q_i(t)$	Queue backlog of the IoT device i in slot t
$D_i(t)$	Amount of computation tasks at IoT device i which will be offloaded in slot t

of IoT device i is denoted by $R_i(t)$ as follows,

$$R_i(t) = B \log_2(1 + \frac{P_i h_i(t)}{B N_0}), \tag{2.1}$$

where B represents the sub-channel's bandwidth, and N_0 is the noise power spectral density.

We study the task offloading of IoT devices in MEC. Define $\boldsymbol{\kappa}(t) = \{\kappa_1(t), \cdots, \kappa_n(t)\}$ as the task offloading decision of IoT devices, where the offloading period of IoT device i is $\kappa_i(t)$. The amount of available offloading task of device i during time slot t are defined as $D_i(t) = R_i \kappa_i(t)$.

Each device keeps a task buffer to save tasks that have generated yet not offloaded. Let $Q_i(t)$ be the task queue backlog of IoT device i at time slot t. Each IoT device cannot offload more tasks than the queue backlog, as given by

$$\kappa_i(t) \le \frac{Q_i(t)}{R_i(t)}, \forall i \in I. \tag{2.2}$$

Because each IoT device operates in a narrowband protocol, it can only access a subchannel at the same time [11]. Therefore, it should satisfy that

$$0 \le \kappa_i(t) \le \tau, \forall i \in I. \tag{2.3}$$

Combining (2.2) and (2.3), we can get that

$$0 \leq \kappa_i(t) \leq T_i(t), \forall i \in I, \tag{2.4}$$

where $T_i(t) = \min\{Q_i(t)/R_i(t), \tau\}$.

According to Lyu et al. [12], we adopt the time division multiple access (TDMA) technology, and different devices can access the same sub-channel at different times in a slot. In addition, in order to offload computing tasks in one slot, IoT devices can access different sub-channels at different times. Therefore, the total offloading time of all IoT devices must not exceed the time slot length of all available sub-channels, as described in (2.5).

$$\sum_{i=1}^{n} \kappa_i(t) \leq S(t)\tau. \tag{2.5}$$

2.1.3 *Task Queuing Model*

We know that $Q_i(t)$ means the task queue backlog of IoT device i at time slot t and $D_i(t)$ indicates the available amount offloading task of device i during time slot t. Then, the evolution of the queue backlog $Q_i(t)$ is derived as,

$$Q_i(t+1) = \max\{Q_i(t) - D_i(t), 0\} + A_i(t). \tag{2.6}$$

In order to ensure the stability of the task queue of these IoT devices and reduce the queuing delay, we bound the average queue backlog of each IoT device across time slots, namely,

$$q_i = \lim_{T \to \infty} \frac{1}{T} \sum_{t=0}^{T-1} \mathbf{E}\{Q_i(t)\} < \varepsilon, \exists\, \varepsilon \in \mathbb{R}^+. \tag{2.7}$$

2.1.4 *Energy Consumption Model*

For each IoT device $i \in I$, the energy consumption of offloading tasks to MEC is affected by two factors, including transmission power and offloading time, i.e., $e_i(t) = P_i\kappa_i(t)$. Therefore, the energy consumed by all IoT devices to transmit data is,

$$e(t) = \sum_{i=1}^{n} P_i\kappa_i(t). \tag{2.8}$$

The quality of the wireless channel is different in different time slots. Therefore, the task offloading speed and transmission energy consumption of IoT devices will also change over time. Therefore, our goal is to obtain the average transmission energy consumption over a relatively long period of time, as shown in (2.9).

$$e = \lim_{T\to\infty} \frac{1}{T} \sum_{t=0}^{T-1} \mathbf{E}\{e(t)\}. \tag{2.9}$$

2.1.5 Problem Statement

The transmission speed of the IoT device when offloading tasks is affected by the wireless channel conditions. Specifically, when the channel conditions are better, the transmission speed is higher. Therefore, when the channel conditions are good, as many tasks as possible should be offloaded to reduce unnecessary time and energy consumption. But if we only choose to offload tasks when the channel conditions are good, this may cause the queue backlog of IoT devices to become very large, and the stability of the queue will be violated. In order to ensure the balance between the offloading energy consumption and the queue, we formulate a unified optimization problem for the computation offloading problem. For this optimization problem, the goal is to minimize the average transmission energy consumption while ensuring the stability of the offloading task queue.

$$\min_{\kappa(t)} \; e = \lim_{T\to\infty} \frac{1}{T} \sum_{t=0}^{T-1} \mathbf{E}\{e(t)\}. \tag{2.10}$$

$$\begin{aligned} s.t. \quad & 0 \le \kappa_i(t) \le T_i(t), \forall i \in I; \\ & \sum_{i=1}^{n} \kappa_i(t) \le S(t)\tau; \\ & q_i = \lim_{T\to\infty} \frac{1}{T} \sum_{t=0}^{T-1} \mathbf{E}\{Q_i(t)\} < \varepsilon, \exists\, \varepsilon \in \mathbb{R}^+. \end{aligned}$$

Remark Problem (2.10) is a stochastic optimization problem. In Problem (2.10), the wireless channel conditions change randomly over time, which are difficult to predict. The optimization goal of Problem (2.10) is to minimize the energy consumption of the system when the wireless channel conditions are constantly changing. Because of the uncertainty of the wireless channel, it is often difficult to obtain statistical information and predict the changing state in the real system. Therefore, it is difficult to solve problem (2.10) offline. Moreover, with the continuous expansion of the scale of the system, the number of IoT devices connected to it will also increase sharply, making the solution space also expanding dramatically, which makes it more difficult to solve the problem. Therefore, an efficient solution needs to be proposed. However, it is very difficult to find a low-complexity and efficient solution to this problem. Therefore, in response to the above challenges, an online task offloading algorithm is proposed in Sect. 2.2, which does not need to know the statistical information of the arrival of the computing task or channel condition.

2.2 EEDCO: Energy Efficient Dynamic Computing Offloading for Mobile Edge Computing

In order to solve the above optimization problem, we design an algorithm EEDOA that can dynamically make offloading decisions. This algorithm can realize the trade-off between task queue and energy consumption without knowing the statistical information of task arrivals or wireless channel quality.

2.2.1 *Joint Optimization of Energy and Queue*

The task queue of IoT devices is expressed as $\Theta(t)$. Then, the Lyapunov function $L(\Theta(t))$ is defined as,

$$L(\Theta(t)) = \frac{1}{2}\sum_{i=1}^{n} Q_i^2(t), \tag{2.11}$$

where $L(\Theta(t))$ represents the queue backlog status of IoT devices. Specifically, $L(\Theta(t))$ increases as the backlog of IoT device queues increases, and decreases as the backlog of IoT device queues decreases. Therefore, in order to maintain a low congestion state of IoT devices, it is necessary to reduce the value of $L(\Theta(t))$.

The conditional Lyapunov drift is defined as,

$$\Delta(\Theta(t)) = \mathbf{E}\{L(\Theta(t+1)) - L(\Theta(t))|\Theta(t)\}. \tag{2.12}$$

It is worth noting that there exists tradeoff between the transmission energy consumption and the queue length. Minimizing the transmission energy consumption of IoT devices may lead to a very large queue length. To this end, we combine transmission energy consumption with queue backlog, and its drift plus energy is,

$$\Delta(\Theta(t)) + V\mathbf{E}\{e(t)|\Theta(t)\}, \tag{2.13}$$

where V is the trade-off parameter between queue stability and transmission energy consumption. We can see that the greater the value of V, the greater the proportion of transmission energy consumption.

Next, we give the upper bound on drift plus energy.

Theorem 2.1 *If the upper bounds of $A_i(t)$ and $R_i(t)$ in each time slot are A_i^{max} and R_i^{max}, then the drift and penalty's upper bound in the task offloading problem is,*

$$\begin{aligned}\Delta(\Theta(t)) + V\mathbf{E}\{e(t)|\Theta(t)\} \leq C \\ + \sum_{i\in I} Q_i(t)\mathbf{E}\{A_i(t) - R_i(t)\kappa_i(t)|\Theta(t)\} \\ + V\sum_{i\in I}\mathbf{E}\{P_i\kappa_i(t)|\Theta(t)\},\end{aligned} \tag{2.14}$$

where $C = \frac{1}{2}[\sum_{i\in I}(A_i^{max})^2 + \sum_{i\in I}(R_i^{max}\tau)^2]$ is a constant.

Proof Using $(\max[a-b,0])^2 \leq a^2 + b^2 - 2ab$ for any $a, b \geq 0$, and Eq. (2.6), we have

$$\begin{aligned}Q_i^2(t+1) \leq Q_i^2(t) + D_i^2(t) + A_i(t)^2 - 2Q_i(t)D_i(t) \\ + 2A_i(t)\max[Q_i(t) - D_i(t), 0].\end{aligned} \tag{2.15}$$

Let $\bar{D}_i(t)$ represent the actual amount of computing tasks to the IoT device i offloaded to MEC server during slot t. Thus, we have

$$\bar{D}_i(t) = \begin{cases} Q_i(t), & Q_i(t) \leq D_i(t) \\ D_i(t), & otherwise. \end{cases} \tag{2.16}$$

Therefore, we can get by (2.16) that $\max[Q_i(t) - D_i(t), 0] = Q_i(t) - \bar{D}_i(t)$. Then, (2.15) is rewritten as,

$$\begin{aligned}Q_i^2(t+1) \leq Q_i^2(t) + D_i^2(t) + A_i^2(t) \\ + 2Q_i(t)[A_i(t) - D_i(t)] - 2\bar{D}_i(t)A_i(t).\end{aligned} \tag{2.17}$$

Since $\bar{D}_i(t)$ and $A_i(t)$ are non-negative, (2.18) holds.

$$\begin{aligned}\frac{1}{2}[Q_i^2(t+1) - Q_i^2(t)] \leq \frac{1}{2}[A_i^2(t) + D_i^2(t)] \\ + Q_i(t)[A_i(t) - D_i(t)].\end{aligned} \tag{2.18}$$

Summing over all the devices on (2.18) and taking conditional expectations, (2.19) holds.

$$\Delta(\Theta(t)) \leq \frac{1}{2}\sum_{i\in I}\mathbf{E}\{A_i^2(t)+D_i^2(t)|\Theta(t)\} + \sum_{i\in I} Q_i(t)\mathbf{E}\{A_i(t)-D_i(t)|\Theta(t)\}. \tag{2.19}$$

Since for any $i \in I$, there holds that $R_i(t) \leq R_i^{max}$ and $\kappa_i(t) \leq \tau$, we obtain

$$D_i(t) = R_i\kappa_i(t) \leq R_i^{max}\tau. \tag{2.20}$$

According to (2.20) and $A_i(t) \leq A_i^{max}$, (2.21) can be obtained.

$$\sum_{i\in I}\mathbf{E}\{A_i^2(t)+D_i^2(t)|\Theta(t)\} \leq \sum_{i\in I}[(A_i^{max})^2+(R_i^{max}\tau)^2]. \tag{2.21}$$

Let C equal to $\frac{1}{2}\sum_{i\in I}[(A_i^{max})^2+(R_i^{max}\tau)^2]$, and add $V\mathbf{E}\{e(t)|\Theta(t)\}$ to (2.19). Thus,

$$\Delta(\Theta(t)) + V\mathbf{E}\{e(t)|\Theta(t)\} \leq C + V\mathbf{E}\{e(t)|\Theta(t)\} + \sum_{i\in I} Q_i(t)\mathbf{E}\{A_i(t)-D_i(t)|\Theta(t)\}. \tag{2.22}$$

Substituting (2.8) into (2.22), it yields (2.14).

2.2.2 *Dynamic Computation Offloading for Mobile Edge Computing*

In this part, to obtain the minimum value of the upper bound of drift plus penalty, we design a dynamic optimization algorithm EEDOA to solve the dynamic calculation offloading. By calculating the upper bound of offloading energy consumption in each time slot, the minimum offloading energy consumption in the whole offloading process is obtained. EEDOA algorithm can effectively reduce the energy consumption of task offloading transmission, maintain the stability of task queue and keep the queue backlog at a low level.

The upper bound of drift plus energy is obtained by EEDOA algorithm, which is expressed as,

$$\min_{\kappa(t)}\{C + \sum_{i\in I} Q_i(t)[A_i(t) - R_i(t)\kappa_i(t)] + V\sum_{i\in I} P_i\kappa_i(t)\}. \tag{2.23}$$

$$s.t. \quad 0 \leq \kappa_i(t) \leq T_i(t), \forall i \in I.$$
$$\sum_{i=1}^{n} \kappa_i(t) \leq S(t)\tau.$$

In each given time slot, because C and $A_i(t)$ are constants, (2.23) can be simplified and rewritten as (2.24).

$$\min_{\kappa(t)} \sum_{i \in I} [V P_i - Q_i(t) R_i(t)] \kappa_i(t). \tag{2.24}$$

For the convenience of calculation, we can transform the above minimization problem into maximization problem, and then get the following maximization problem,

$$\max_{\kappa(t)} \sum_{i \in I} \omega_i(t) \kappa_i(t). \tag{2.25}$$
$$s.t. \quad 0 \leq \kappa_i(t) \leq T_i(t), \forall i \in I.$$
$$\sum_{i=1}^{n} \kappa_i(t) \leq S(t)\tau.$$

In (2.25), $\omega_i(t) = Q_i(t) R_i(t) - V P_i$.

Problem (2.25) can be understood as that given a group of users, each user makes its own decision to optimize the total benefit value. This is similar to the combinatorial optimization of knapsack problem. It can be regarded as a special knapsack problem. In this knapsack problem, the total capacity of the knapsack is $S(t)\tau$ and the unit value of each item is $\omega_i(t)$. Before solving this special knapsack problem, we first sort all IoT devices according to the non increasing order of $\omega_i(t)$. The rule of loading the backpack is that the greater the $\omega_i(t)$ value of the Internet of things device, the earlier it is loaded into the backpack. The arbitrary optimal solution is to select the item with the largest nonnegative $\omega_i(t)$ to load into the backpack. Finally, when the remaining capacity of the backpack is empty or the $\omega_i(t)$ value of the selected IoT is not positive, the loading is terminated.Therefore, the index of interrupt item χ can be deduced as,

$$\chi = \min\{\chi_1, \chi_2\}, \tag{2.26}$$

where

$$\chi_1 = \arg\min_{i} \sum_{j=1}^{i} T_i(t) > S(t)\tau, \tag{2.27}$$

Algorithm 1 Energy efficient dynamic offloading algorithm (EEDOA)

1: **for** $i \in I$ **do**
2: Calculate the $R_i(t)$, $T_i(t)$ and $\omega_i(t)$.
3: **end for**
4: Sort all the devices i in the descending order of $\omega_i(t)$.
5: Set the index χ according to (2.26).
6: **for** $i \in I$ **do**
7: Set the offloading decision $\kappa_i(t)$ according to (2.29).
8: **end for**

$$\chi_2 = \arg\max_i \ \omega_i(t) \geq 0. \tag{2.28}$$

Then, the optimal offloading decision is given as,

$$\kappa_i^*(t) = \begin{cases} T_i(t), & i < \chi \\ \min\{S(t)\tau - \sum_{i=1}^{\chi-1} T_i(t), T_\chi(t)\}, & i = \chi \\ 0, & i > \chi. \end{cases} \tag{2.29}$$

Remark As mentioned earlier, for each IoT device, the EEDOA algorithm defines its unit offload profit as $\omega_i(t)$. We know that there is a trade-off between the queue backlog on IoT devices and the transmission energy consumption when tasks are offloaded. Thus, $\omega_i(t)$ depends on the current queue backlog and channel congestion. Therefore, in order to minimize the offloading transmission energy consumption and maintain the stability of the task queue at the same time, in each time slot t, the IoT device with the largest non-negative $\omega_i(t)$ is preferentially selected for offloading tasks. In addition, in order to perform task offloading more flexibly, the value of V can be adjusted to achieve a balance between transmission energy consumption and queue backlog.

Algorithm 1 presents the details of the EEDOA algorithm.

2.2.3 *Trade-Off Between Queue Backlog and Energy Efficiency*

In this part, we analyze the relationship between queue length and energy consumption. When the IoT device offloads tasks, on the one hand, it will cause energy consumption, on the other hand, it will also cause queue backlog. The IoT device will be affected by channel conditions and its own transmission power. When the channel conditions are good, the transmission speed increases and the number of tasks that can be offloaded will also increase, which will reduce the length of the task queue. But the number of offloaded tasks increases, more energy consumption will also be caused. When the channel conditions are bad, the number of tasks that can be offloaded becomes smaller, which reduces the offloading energy consumption, but makes the backlog of tasks on the IoT devices larger, increasing the burden of the IoT devices. Therefore, we need to balance offloading energy consumption and task backlog. However, the task offloading process of IoT devices changes dynamically and it is difficult for us to accurately obtain the timely information of each IoT device.

Because we focus on dynamic computation offloading in mobile edge computing, for this random optimization problem, we can use stochastic optimization techniques to ensure the stability of task queue. We can achieve the balance between energy consumption and queue backlog according to Lyapunov drift plus penalty function, and the system is in equilibrium by adjusting the value of parameter V.

2.2.4 Convergence and Complexity Analysis

In this part, we use the mathematical method to analyze the performance of EEDOA, obtain the upper bound of offloading energy consumption and task backlog. Then, the complexity of EEDOA is analyzed, and the worst case time complexity is given.

The average time queue backlog of IoT devices is given below, which is defined as $\bar{Q}$.

$$\bar{Q} = \lim_{T \to \infty} \frac{1}{T} \sum_{t=0}^{T-1} \sum_{i=1}^{n} \mathbf{E}\{Q_i(t)\}. \tag{2.30}$$

In order to obtain the optimal transmission energy efficiency, the offloading decision $\kappa(t)$ is made according to the fixed probability distribution. Lemma 2.1 is given as follows.

Lemma 2.1 *For any offloading task arrival rate* $\boldsymbol{\lambda}$*, there is a task offloading decision* $\boldsymbol{\pi}^*$*, which does not depend on the current task queue backlog and satisfies,*

$$\mathbf{E}\{e^{\pi^*}(t)\} = e^*(\boldsymbol{\lambda}),$$

$$\mathbf{E}\{A_i(t)\} \leq \mathbf{E}\{R_i(t)\kappa_i^{\pi^*}(t)\},$$

where $\boldsymbol{\Lambda}$ *indicates the capacity of the system to admit offloading tasks, and* $e^*(\boldsymbol{\lambda})$ *represents the minimum time average offloading transmission energy consumption under* $\boldsymbol{\Lambda}$.

Proof Lemma 2.1 can be proved by Caratheodory's theorem. Here, for the sake of simplicity and convenience for readers, we omit the detailed proof. □

Recall that we assume that there is an upper bound $A_i(t)$ for the task arrival amount of each IoT device in each time slot t. Thus, there is also a bound for our transmission energy consumption. And the upper bound of energy consumption is $\hat{e}$, and the lower bound of energy consumption is $\check{e}$. Then, based on Lemma 2.1, we can get the following Theorem 2.2. In Theorem 2.2, we give the upper bound of transmission energy consumption and task queue length.

Theorem 2.2 *It is assumed that ϵ exists and satisfies $\boldsymbol{\lambda} + \epsilon \in \boldsymbol{\Lambda}$. For the given V, the transmission energy consumption of EEDOA satisfies,*

$$e^{EEDOA} \le e^* + \frac{C}{V}. \tag{2.31}$$

At the same time, we get the average queue length in (2.32).

$$\bar{Q} \le \frac{C + V(\hat{e} - \check{e})}{\epsilon}, \tag{2.32}$$

where C is the constant given by Theorem 2.1, and e^* represents the optimal transmission energy consumption with $\boldsymbol{\lambda}$.

Proof For a given task arrival rate $\boldsymbol{\lambda} + \epsilon$, according to Lemma 2.1, there is a randomization strategy π', and it satisfies,

$$\mathbf{E}\{e^{\pi'}(t)\} = e^*(\boldsymbol{\lambda} + \epsilon), \tag{2.33}$$

$$\mathbf{E}\{A_i(t)\} + \epsilon \le \mathbf{E}\{R_i(t)\kappa_i^{\pi'}(t)\}. \tag{2.34}$$

□

Because (2.22)'s R.H.S is minimized in the EEDOA algorithm, for the offloading decision, we can get

$$\begin{aligned}\Delta(\Theta(t)) + V\mathbf{E}\{e(t)|\Theta(t)\} &\le C + V\mathbf{E}\{e^{\pi'}(t)|\Theta(t)\} \\ &+ \sum_{i\in I} Q_i(t)\mathbf{E}\{A_i(t) - R_i(t)\kappa_i^{\pi'}(t)|\Theta(t)\}.\end{aligned} \tag{2.35}$$

Combining (2.33), (2.34) and (2.35), we can get

$$\Delta(\Theta(t)) + V\mathbf{E}\{e(t)|\Theta(t)\} \le C + Ve^*(\boldsymbol{\lambda} + \epsilon) - \epsilon \sum_{i\in I} Q_i(t). \tag{2.36}$$

By taking expectations on both sides of Eq. (2.36) at the same time, we have

$$\begin{aligned}&\mathbf{E}\{L(\Theta(t+1)) - L(\Theta(t))\} + V\mathbf{E}\{e(t)\} \\ &\quad \le C + Ve^*(\boldsymbol{\lambda} + \epsilon) - \epsilon \sum_{i\in I} \mathbf{E}\{Q_i(t)\}.\end{aligned} \tag{2.37}$$

For both sides of Eq. (2.37), start from time slot 0 and add to time slot $t-1$ at the same time, we can get

$$\begin{aligned}&\mathbf{E}\{L(\Theta(T))\}-\mathbf{E}\{L(\Theta(0))\}+V\sum_{t=0}^{T-1}\mathbf{E}\{e(t)\}\\&\leq CT+VTe^*(\boldsymbol{\lambda}+\epsilon)-\epsilon\sum_{t=0}^{T-1}\sum_{i\in I}\mathbf{E}\{Q_i(t)\}.\end{aligned}\tag{2.38}$$

Generally speaking, we suppose that the task queue length of IoT devices is empty when $t=0$, that is, $L(\Theta(0))=0$. Thus, we have,

$$V\sum_{t=0}^{T-1}\mathbf{E}\{e(t)\}\leq CT+VTe^*(\boldsymbol{\lambda}+\epsilon)-\epsilon\sum_{t=0}^{T-1}\sum_{i\in I}\mathbf{E}\{Q_i(t)\}.\tag{2.39}$$

Because $Q_i(t)$ and ϵ are non-negative, we can get (2.40).

$$V\sum_{t=0}^{T-1}\mathbf{E}\{e(t)\}\leq CT+VTe^*(\boldsymbol{\lambda}+\epsilon).\tag{2.40}$$

Divide both sides of Eq. (2.40) by VT at the same time, it holds

$$\frac{1}{T}\sum_{t=0}^{T-1}\mathbf{E}\{e(t)\}\leq\frac{C}{V}+e^*(\boldsymbol{\lambda}+\epsilon).\tag{2.41}$$

When $\epsilon\to 0$, $T\to\infty$, apply the Lebesgues dominated convergence theorem, we obtain (2.31).

According to (2.39), we also obtain

$$\begin{aligned}&\epsilon\sum_{t=0}^{T-1}\sum_{i\in I}\mathbf{E}\{Q_i(t)\}\\&\leq CT+VTe^*(\boldsymbol{\lambda}+\epsilon)-V\sum_{t=0}^{T-1}\mathbf{E}\{e(t)\}\\&\leq CT+VT(\hat{e}-\check{e}).\end{aligned}\tag{2.42}$$

Divide both sides of Eq. (2.42) by ϵT at the same time, and letting $T\to\infty$, we obtain (2.32).

Remark In Theorem 2.2, the upper bound of transmission energy consumption and the upper bound of queue backlog are given. In Eq. (2.31), the transmission energy consumption decreases as V increases. In Eq. (2.32), the queue length increases as V increases. Combining Theorem 2.2 and (2.31), we can see that our algorithm achieves a $[O(1/V), O(V)]$ balance between transmission energy consumption and queue length. It can be seen that in order to achieve the optimal transmission energy consumption, V can be set to be large enough, but this will cause a large backlog of task queues. Therefore, we can set a suitable V value to achieve the balance between the energy consumption and the queue backlog that we need.

Then, we analyze the complexity of the EEDOA algorithm. There are two loops (lines 1–3, lines 6–8). In the EEDOA algorithm, each IoT device is traversed once. Therefore, the operation complexity of each cycle is $O(n)$, where n represents the number of IoT devices. In the fourth row of the algorithm, we use the quick sort method to sort the IoT devices in descending order of W. In line 7, because the offloading decisions of different IoT devices do not affect each other, the complexity of the second loop is $O(n)$. In summary, the time complexity of the EEDOA algorithm is $O(n)$.

2.3 Performance Evaluation

In this part, we evaluate our EEDOA algorithm. We test our EEDOA algorithm according to different parameters. Then, we conduct comparative experiments. The effectiveness of our proposed EEDOA algorithm is verified by comparing with two benchmark algorithms.

In the experiment, we consider a network consisting of 100 IoT devices and an edge server. The IoT device offloads computing tasks to the edge server, and the length of the time slot is $\tau = 1$ s. The amount of data arriving per time slot on the IoT device i meets the uniform distribution, i.e., $A_i(t) \sim U[0, 2200]$ bits. The wireless channel is considered as a small-scale Rayleigh fading model, and the channel power gain $h_i(t)$ follows the exponential distribution of the unit mean, i.e., $h_i(t) \sim E(1)$ [13]. We set the number of available sub-channels as $S(t)$, i.e., $S(t) \sim U[10, 30]$. The transmit power of each IoT device is $P_i \sim U[10, 200]$ mW [12]. The bandwidth of the sub-channel is set as $B = 1$ MHz. The noise power spectral density is set as $N_0 = 10^{-6}$ W/Hz. The parameter settings are summarized in Table 2.2. To improve the reliability of the experiments, each setting of the experiments is run 3000 times, and then the average value is taken.

2.3.1 *Impacts of Parameters*

2.3.1.1 Effect of Tradeoff Parameter

Figures 2.1 and 2.2 show the changes in transmission energy consumption and queue length as the parameter V changes. In the EEDOA algorithm, the parameter V is used to weigh the queue length and transmission energy consumption.

Table 2.2 Simulation parameters

Parameters	Values
Time slot length	1 s
The amount of data arriving at IoT device i per time slot	$U[0, 2200]$ (bit)
The channel power gain	$E(1)$
The number of available sub-channels	$U[10, 30]$
The transmit power of each IoT device	$U[10, 200]$ (mW)
The bandwidth of the sub-channel	1 MHz
The noise power spectral density	10^{-6} W/Hz

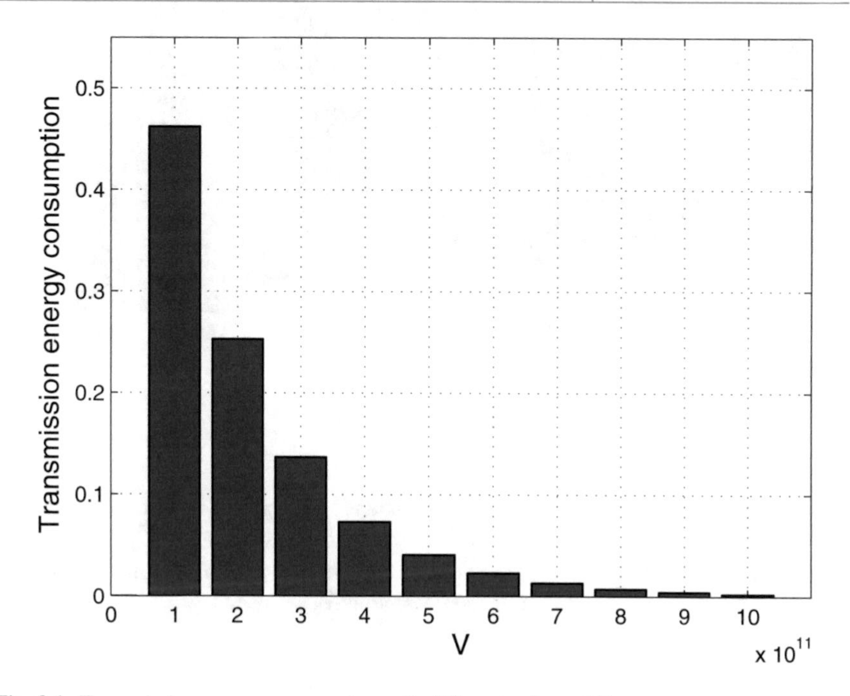

Fig. 2.1 Transmission energy consumption with different values of V

Figure 2.1 shows the change of transmission energy consumption under the influence of parameter V. It can be seen from the figure that the transmission energy consumption is negatively correlated with the parameter V. As V increases, the transmission energy consumption decreases. To explore the reason, as V increases, the weight of transmission energy consumption also increases, and the MEC system pays more attention to energy saving. The EEDOA algorithm reduces the total energy consumption of the system by dynamically adjusting task offloading decisions. The result in Fig. 2.1 conforms to Eq. (2.31) in Theorem 2.2.

Figure 2.2 shows how the queue length changes under the influence of the parameter V. It can be seen from the figure that the queue length increases as V increases. However, the queue length does not increase indefinitely, and will

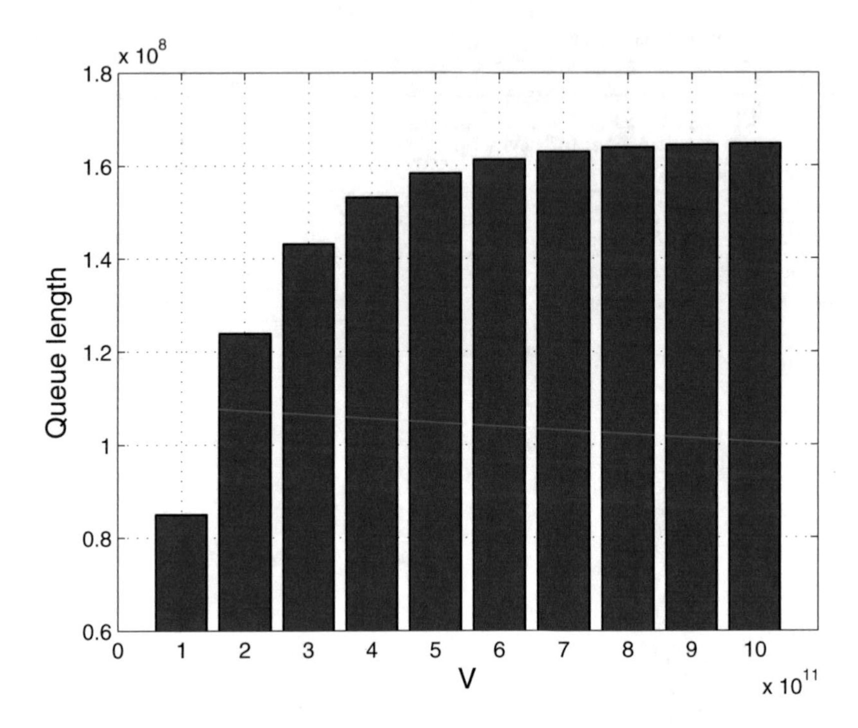

Fig. 2.2 Queue length with different values of V

gradually converge, eventually reaching a balance. The result in Fig. 2.2 conforms to Eq. (2.32) in Theorem 2.2. As can be seen from Figs. 2.1 and 2.2, the EEDOA algorithm can reduce transmission energy consumption and ensure the stability of the queue by changing the value of the parameter V.

2.3.1.2 Effect of Arrival Rate

Figures 2.3 and 2.4 show the impact of different task arrival rates on transmission energy consumption and queue length. In this experiment, we set that the task arrival rate of each IoT device as $\alpha \cdot A_i(t)$, and the value of α is 0.8, 1 and 1.2.

Figure 2.3 shows the impact of task arrival rate on transmission energy consumption. It can be seen from the figure that for different task arrival rates, the transmission energy consumption all increases rapidly at first and then gradually stabilizes. However, at the same time slot, the higher the task arrival rate, the higher the transmission energy consumption. This is because the higher the task arrival rate, the more tasks arrive at each time slot. Since the local processing capacity under different task arrival rates remains unchanged, when the task arrival rate is higher, the more tasks are offloaded and the more transmission energy consumption. After reaching the processing limit of the server, the EEDOA algorithm can dynamically

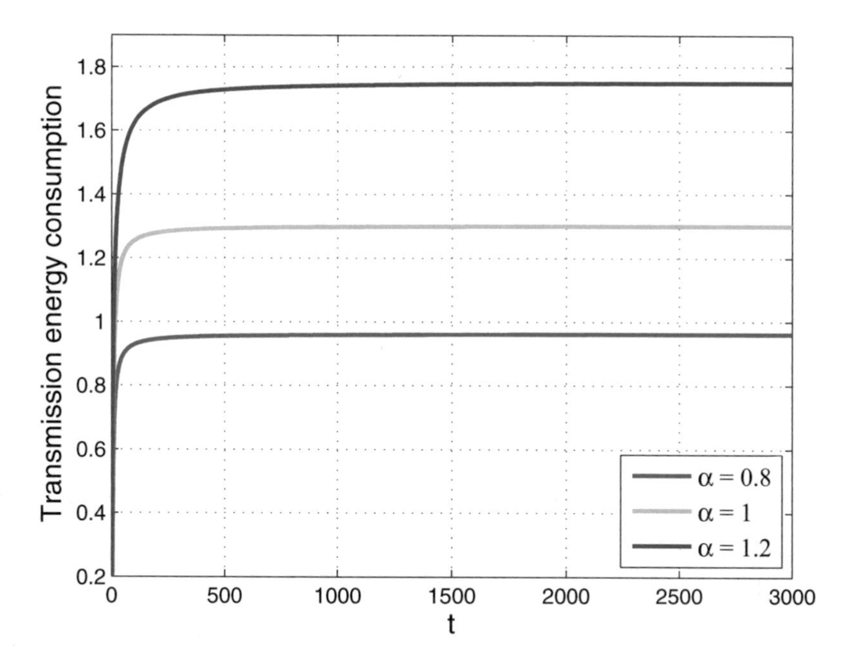

Fig. 2.3 Transmission energy consumption with different arrival rates

adjust the offloading decision. In this way, the transmission energy consumption will remain stable.

Figure 2.4 shows the effect of task arrival rate on queue length. It can be seen from the figure that for different task arrival rates, the queue length all increases rapidly at first and then gradually stabilizes. However, at the same time slot, the higher the task arrival rate, the higher the queue length. This is because the higher the task arrival rate, the more tasks arrive at each time. Because the servers have the same processing power under different task arrival rates, when the task arrival rate is higher, the more tasks are backlogged and the larger queue length. It can be seen from the figure that although the task arrival rate is different, the queue length will eventually converge. This result shows that EEODA can dynamically adjust the offloading decision according to the different arrival rate, and make the system stabilize quickly.

2.3.1.3 Effect of Transmit Power

Figures 2.5 and 2.6 show the effect of different transmit powers on transmission energy consumption and queue length. In the experiment, we set the transmit power of each IoT device as $\beta \cdot P_i$, and the value of β is 0.8, 1, and 1.2.

Figure 2.5 shows the relationship between transmit power and transmission energy consumption. It can be seen from the figure that for different transmit

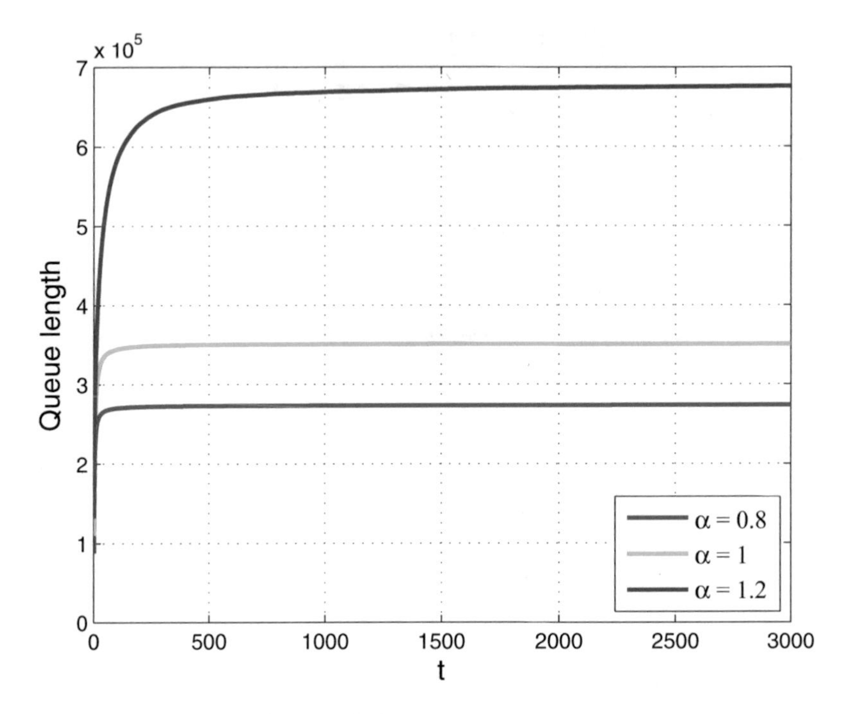

Fig. 2.4 Queue length with different arrival rates

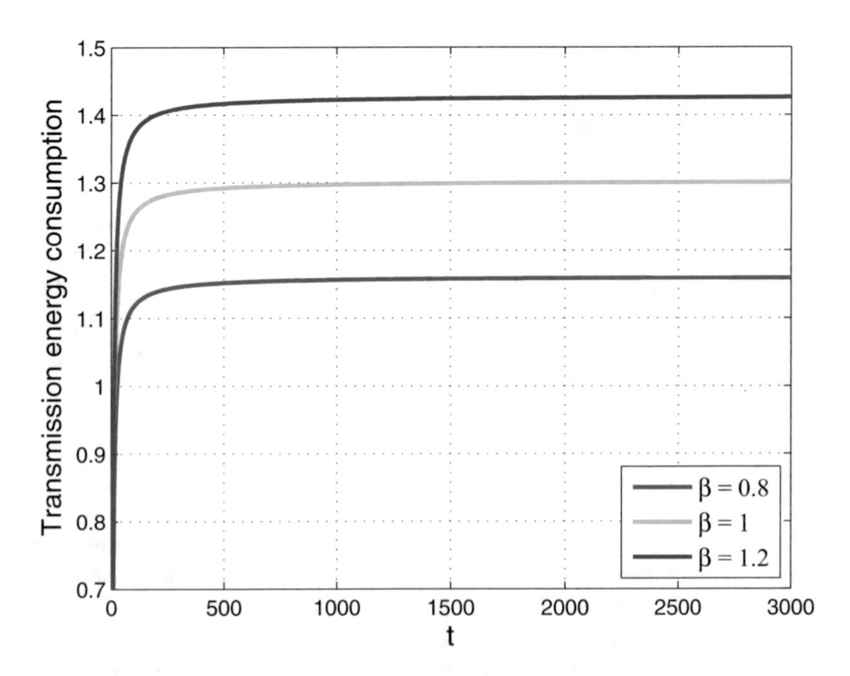

Fig. 2.5 Transmission energy consumption with different transmit powers

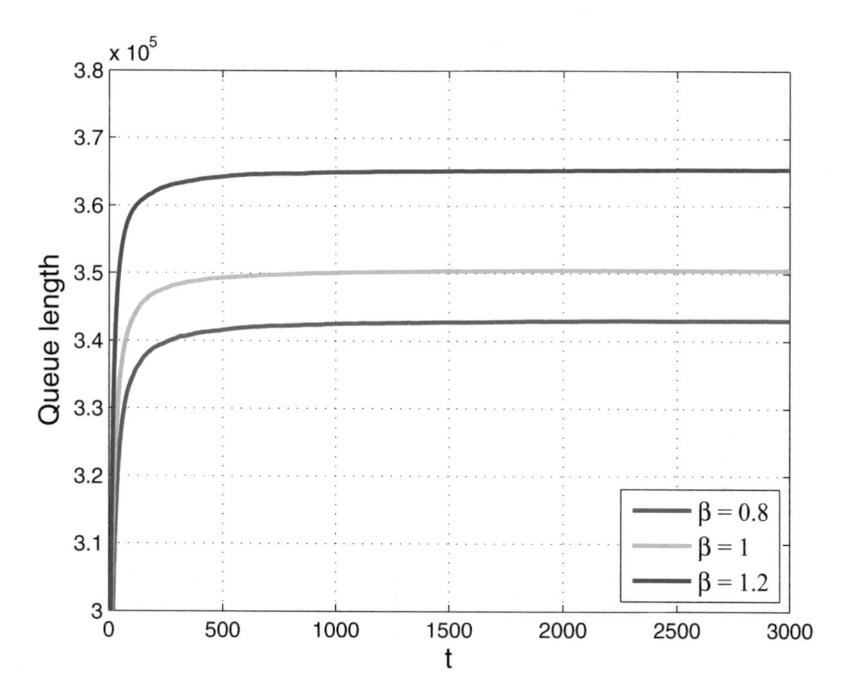

Fig. 2.6 Queue length with different transmit powers

powers, transmission energy consumption first increases rapidly, and then gradually stabilizes. However, in the same time slot, the higher the transmission power, the higher the transmission energy consumption. This is because the higher the transmit power, the higher the corresponding transmission energy consumption when each IoT device transmits data. Therefore, the total energy consumption of the system will also increase.

Figure 2.6 shows the relationship between transmit power and queue length. It can be seen from the figure that for different transmit powers, the queue length increases rapidly first, and then gradually stabilizes. However, in the same time slot, the greater the transmit power, the greater the corresponding queue length. This is because the processing capacity of the server is limited. When the transmit power increases, in order to reduce the transmission energy consumption, the EEDOA algorithm will dynamically reduce the amount of tasks offloaded by the IoT device, thereby increasing the queue length. It can be seen from the figure that under different transmission powers, the queue length will eventually stabilize gradually, which also meets our requirements.

2.3.1.4 Effect of Channel Power Gain

Figures 2.7 and 2.8 show the influence of different channel power gains on transmission energy consumption and queue length. In our experiment, we set that the channel power gain is $\gamma \cdot \mu$, and the value of γ is 0.8, 1 and 1.2.

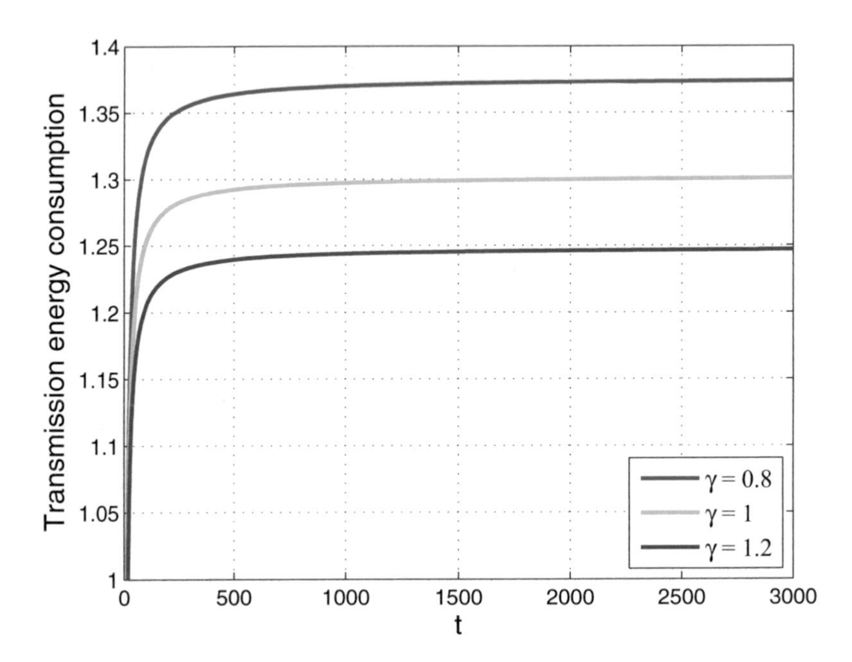

Fig. 2.7 Transmission energy consumption with different channel power gains

Figure 2.7 shows the relationship between channel power gain and transmission energy consumption. It can be seen from the figure that under different channel power gains, the transmission energy consumption first increases rapidly, and then gradually stabilizes. However, in the same time slot, the higher the channel power gain, the lower the transmission energy consumption. This is because when the channel power gain is higher, the higher the transmission rate. Then it takes less time for IoT devices to offload the same amount of data, and the transmission energy consumption is also less.

Figure 2.8 shows the relationship between channel power gain and queue length. It can be seen from the figure that in the same time slot, the higher the channel power gain, the smaller the queue length. This is because the higher the channel power gain, the higher the offloading rate. Then, IoT devices will offload more data per time slot, and the queue length is smaller.

2.3.1.5 Effect of Number of IoT Devices

Figures 2.9 and 2.10 show the impact of the number of different IoT devices on transmission energy consumption and queue length. The number of IoT devices ranges from 70 to 100, with an increase of 10.

Figure 2.9 shows the relationship between the number of IoT devices and transmission energy consumption. As can be seen from the figure, transmission

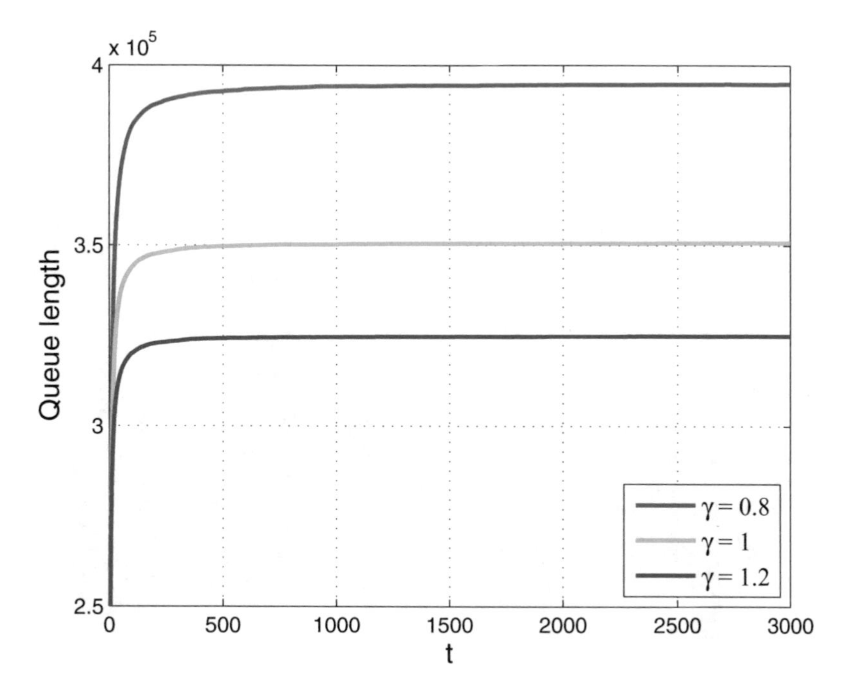

Fig. 2.8 Queue length with different channel power gains

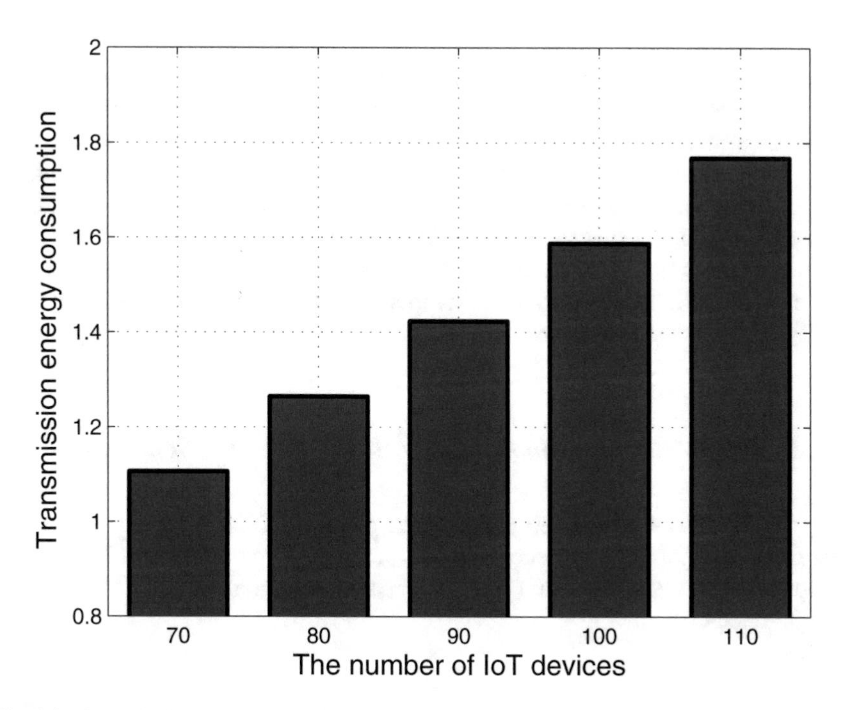

Fig. 2.9 Transmission energy consumption with different numbers of IoT devices

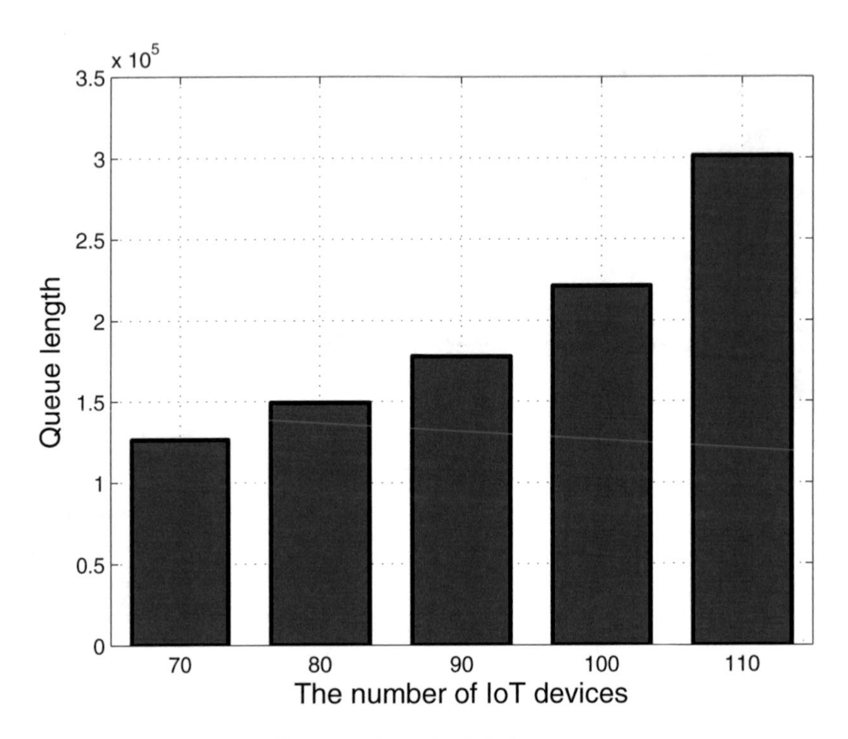

Fig. 2.10 Queue length with different numbers of IoT devices

energy consumption increases with the increase of the number of IoT devices. Because when the number of IoT devices increases, the total number of offloading tasks will also increase, resulting in an increase in transmission energy consumption.

Figure 2.10 shows the relationship between the number of IoT devices and the queue length. As can be seen from the figure, the queue length increases with the increase of the number of IoT devices. The reason is that the more IoT devices, the more computing tasks will be offloaded. However, the total transmission rate of the system is limited, thereby only some computing tasks can be offloaded, and the rest will be stored in the task buffer. This leads to a rapid increase in queue length.

2.3.2 *Performance Comparison with EA and QW Schemes*

In order to better evaluate the performance of the EEDOA algorithm and prove that EEDOA algorithm can better solve the problem of computation offloading, we compare the EEDOA algorithm with the equal allocation algorithm and the queue weighting algorithm.

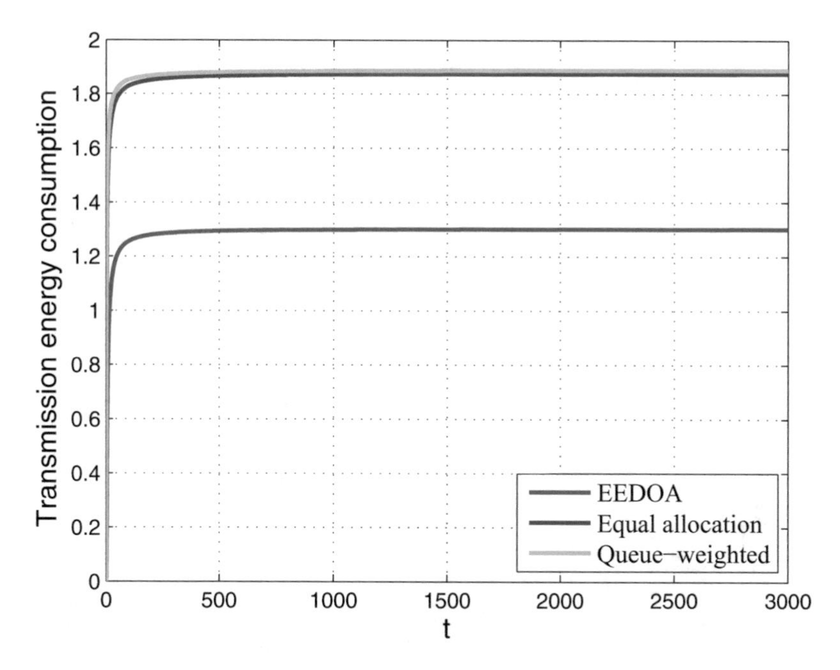

Fig. 2.11 Transmission energy consumption with three different algorithms

- **Equal Allocation (EA)**: This algorithm divides the time of a time slot t into equal parts according to the number of IoT devices, and each IoT device is allocated the same offload transmission time.
- **Queue-Weighted (QW)**: This algorithm allocates offloading time in a weighted manner. In each time slot t, the task offload time of each IoT device is allocated according to its respective weighted queue length. The principle of offloading time allocation is that the larger the weighted queue length of the IoT device, which indicates that it has a higher priority, more offloading time is allocated to it. On the contrary, the less offloading time is allocated to it.

Figure 2.11 shows the relationship between the transmission energy consumption of three different algorithms. Figure 2.12 shows the relationship between the queue length of the three different algorithms. It can be seen from the figure that our EEDOA algorithm can minimize transmission energy consumption and queue length. This experiment proves the effectiveness of our algorithm. It is proved that our algorithm can make the better choice in reducing transmission energy consumption and reducing queue length. The reason is as follows. Our algorithm can adapt to the dynamic changes of the channel. That is, EEDOA algorithm can detect the change of channel state in time and respond quickly. Then dynamically adjust the offloading decision of IoT devices to maintain the balance between queue length and transmission energy consumption.

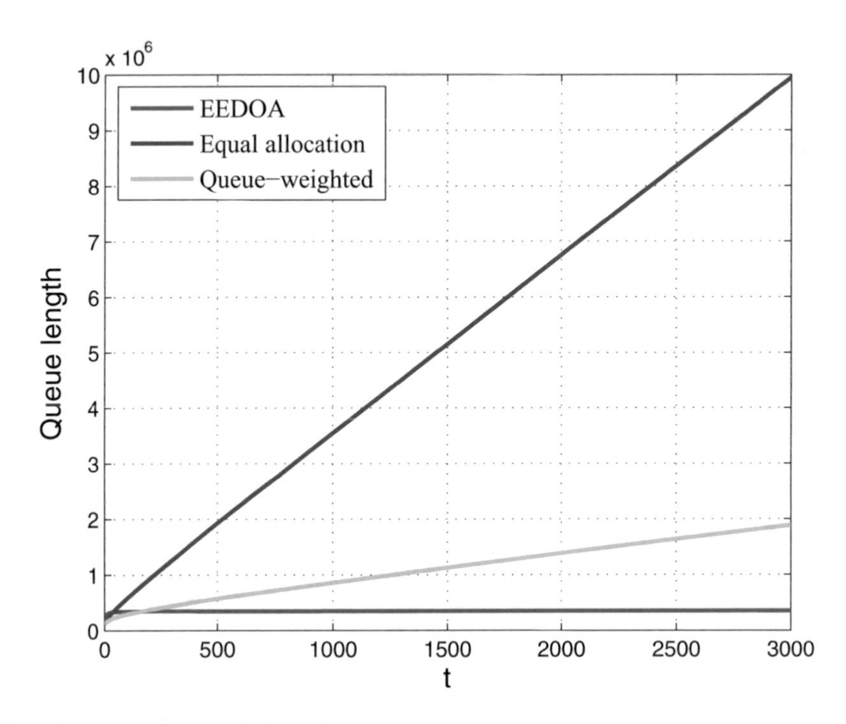

Fig. 2.12 Queue length with three different algorithms

From Fig. 2.11, we can see that the transmission energy consumption of the Equal allocation strategy and the Queue-weighted strategy are very close. However, from Fig. 2.12, we can see that the queue length of the Equal allocation strategy is greater than the Queue-weighted strategy. This is because in the Queue-weighted strategy, IoT devices with a longer queue length will be allocated more offloading time, which can offload more backlog tasks, thereby reducing the total queue length of all IoT devices. However, the Queue-weighted strategy does not consider the channel state, which is different from our algorithm. Our EEDOA algorithm considers the channel state, thereby the queue length of our algorithm is lower than the Queue-weighted strategy. From Figs. 2.11 and 2.12, we can see that the EEDOA algorithm has great advantages in reducing transmission energy consumption and queue length, and can better balance transmission energy consumption and queue length.

2.4 Literature Review

In recent years, with the development of IoT technology, MEC computing mode has attracted more and more attention. The problem of computing task offloading based on MEC has been studied a lot. Munoz et al. [10] proposed femto cloud based on femto access point, which could reduce the battery loss and delay of mobile devices by providing computing power for femto access point. Then, they proposed a framework to jointly optimize the use of radio and computing resources by considering the tradeoff between energy consumption and delay. Wang et al. [14] proposed MCC based C-RAN combined with cloud wireless access network and MCC, and studied the joint energy minimization and resource allocation of C-RAN with MCC under the time constraint of a given task. Then, according to the time model of computing and communication, a non convex energy minimization problem was proposed. Then, according to the constraints of the problem, it was transformed into an equivalent convex problem based on weighted least mean square error, and the problem was solved by iterative algorithm.

Chen et al. [15] studied the problem of multi-user computation offloading for mobile edge cloud computing in a multi-channel wireless interference environment, and proved that the optimal solution of computing concentration was NP-hard. Then, they proposed an algorithm for computation offloading using game theory. And it was proved through experiments that the algorithm could obtain good computation offloading performance even when the user scale continued to increase. Yu et al. [16] believed that optimizing computing or wireless resource allocation alone was highly suboptimal. Because they believed that the congestion of computing resources would lead to the waste of wireless resources. Thus, they studied the resource allocation strategy of MEC system, namely, MEC could be accessed by multiple devices at the same time through orthogonal frequency division multiplexing. At the same time, a joint scheduling algorithm for coordinating the allocation of wireless resources and computing resources was proposed, which could save energy while processing more offloading requests.

Lyu et al. [17] studied computation offloading on proximate cloud. According to the variability of mobile device functions and user preferences, a heuristic computation offloading algorithm was proposed. The system utility was maximized by jointly optimizing computation offloading decision, communication resources and computing resources. They solved the problem of computing and communication resource optimization through convex optimization method. It was proved that this problem was NP-hard. In order to extend the battery life of mobile devices, You et al. [18] proposed a solution integrating MCC and microwave power transmission technology. Then, they proposed an energy-saving computing framework, which could maximize the probability of successfully computing a given data under given energy collection and time constraints. The optimization problem was transformed into an equivalent problem of minimizing local computing energy consumption and offloading energy consumption, which was solved by convex optimization theory.

Chen et al. [19] studied the dynamic resource management problem of MEC in the IoT. This problem combined power control and computing resource management. They used Markov decision process to minimize the average delay of tasks, and proposed a dynamic resource management algorithm based on deep reinforcement learning to solve the Markov decision process problem. In the algorithm, a deep deterministic strategy gradient was used, it could effectively reduce the average delay of tasks. Zhang et al. [20] studied the task scheduling problem to solve the task offloading problem of IoT devices under multiple servers. The task scheduling problem was modeled as an optimization problem, and the optimization problem was proved to be NP-hard. Thereby, a two-stage task scheduling cost optimization algorithm was proposed. Through this algorithm, the computation overhead of the system could be effectively reduced while meeting all task delay requirements.

However, these studies were carried out on the premise that the task arrival and channel state could be predicted. Due to the complexity of the system, the offloading task of IoT devices will be affected by these factors such as channel congestion and interference between IoT devices. It is impossible to predict or know the offloading amount in advance. And the state of the channel is also highly dynamic, which is difficult to predict accurately, thus it is unrealistic to solve the problem in this way. In order to solve the above-mentioned problem, some optimization techniques have been introduced and have been applied in some research work. Mao et al. [13] studied an MEC system with an energy harvesting device that could power mobile devices through energy harvesting technology, and proposed a low-complexity dynamic computation offloading algorithm based on Lyapunov optimization. This decision only depended on the current state of the system, and the minimum execution cost of the system could be given without calculating the task request, wireless channel and energy acquisition process information.

Lyu et al. [21] proposed a distributed online fog computing optimization method, which could suppress the selfish behavior of devices with a tooth-for-tooth mechanism. The method encouraged devices to cooperate in order to minimize tolerable delay and average time energy consumption. They proposed a low-complexity algorithm implemented in a completely distributed manner to optimize energy consumption. Kwak et al. [22] studied the computation offloading problem in the single-user MEC system, and jointly considered three dynamic problems in the real mobile environment. They proposed an algorithm by using Lyapunov optimization, which could be used in a given delay minimize energy consumption under constraints.

Jiang et al. [23] proposed a cloud offload scheduling scheme based on Lyapunov optimization, which was suitable for multiple applications on multi-core mobile devices. They proposed an online algorithm, which could balance the average power consumption and the average queue length of cores. But the main research content of the above-mentioned work was based on the single-user system. Because the multi-user system involves the mutual restriction between users, it is more difficult to design a good multi-user system offloading strategy.

However, with the development of IoT technology, the number of IoT devices is increasing rapidly, and more IoT devices will generate more offloading tasks, which will bring greater burden to the system. Therefore, designing a suitable task offloading algorithm for low-complexity multi-user systems is very important.

Lyu et al. [12] studied the task scheduling problem and proposed a scheduling method that could tolerate outdated network knowledge. Based on this, a perturbed Lyapunov function was proposed, which could maximize the average time network effect, thereby ensured the balance between throughput and fairness. By transforming the deterministic optimization of each time slot into linear relaxation of knapsack problem, the optimal offloading strategy was obtained. Mao et al. [13] studied the resource management of the multi-user MEC system and used the Gauss-Seidel method to determine the optimal transmit power and bandwidth allocation. Then, they proposed an improvement mechanism, and established an optimization model to achieve the balance between queue stability and power optimization. Chen et al. [24] studied the problem of joint task scheduling and energy management, and established a stochastic optimization problem constrained by queue stability and energy budget. At the same time, the calculation throughput and fairness between different units were considered, and the optimization problem was transformed into a deterministic problem. Then, it was decoupled into four independent subproblems that does not require future system statistical information to achieve a balance between system utility and queue.

Chen et al. [25] studied the efficient request scheduling and service management of large-scale computing systems. At the same time, they also considered the restrictive relationship between performance, queue congestion and energy consumption. Based on this, they proposed a distributed online scheduling management algorithm. The algorithm achieved a trade-off between system stability and energy consumption optimization without prior knowledge of the statistics of arriving requests. Chen et al. [26] studied the request scheduling and resource allocation in the multi cloud environment of the IoT. Because in the multi cloud, each cloud had its own virtual machine type and pricing scheme, the author formulated a stochastic optimization problem to capture the dynamics in the multi cloud environment. Then, they applied the stochastic optimization theory to minimize the system cost and queue delay of IoT applications. Thereby, the tradeoff between system cost and queue delay was realized. Chen et al. [27] studied the dynamic computation offloading problem in the edge computing of the IoT, that is, to achieve the balance between offloading cost and performance when the task generation process was highly dynamic. They proposed a dynamic computation offloading algorithm, which decomposed the optimization problem into multiple subproblems, and then solved these subproblems in an online distributed way.

Chen et al. [28] proposed a distributed online web service selection method, which jointly considered the response time, queue congestion and energy consumption. Moreover, this method does not need to know a prior statistical information of task arrival in advance, and could optimize the energy efficiency while stabilizing the system. Chen et al. [29] studied the problem of search space reduction in multi-objective optimization, and introduced the Pareto set of QoS-aware service composition. Then, they proposed a service composition method based on partial selection technology, which could achieve greater performance improvement. Chen et al. [30] proposed an IoT device management and service composition method based on social network. According to the relationship between IoT devices and fully considering social attributes, they designed a set of IoT service composition scheme that could meet the needs of users in many aspects.

Chen et al. [31] studied the MEC-based task offloading and channel resource allocation problems in ultra-dense networks, and defined the task offloading problem as an integer nonlinear programming problem. Due to the coupling between the decision variables, they also proposed a differential evolution-based method that could minimize the system energy consumption while meeting the maximum calculation delay. Chen et al. [32] considered the scenario with dense distribution of edge nodes. They studied the offloading problem of overlapping coverage areas under multiple edge servers and used stochastic game theory to establish a two-step game model. By proving the existence of Nash equilibrium, the offloading costs could be significantly reduced.

Yao et al. [33] studied the task offloading problem under time-varying conditions. Considering that the data generated by each user had different execution priorities, they proposed a multi-user computation offloading framework based on mobile edge computing to solve the scheduling problem in task queues. The framework used Deep Q Network to solve the optimal offloading strategy, and the tasks with high priority would be processed first, which was more consistent with the actual scene in real life. Yang et al. [34] studied the green-oriented problem in the MEC network and defined the green-oriented problem as a delay-constrained energy minimization problem, that is, to minimize the energy consumption of all users in the MEC system. They believed that the problem of finding the optimal strategy was a mixed integer linear programming problem. In order to solve this NP-hard problem, reinforcement learning was used to solve it. And long-term goals were considered, which could achieve good energy saving goal.

Zhao et al. [35] studied the utilization of computing resources between IoT devices and edge servers. In order to ensure the QoS of IoT devices, they used energy harvesting devices to minimize energy consumption. Then used Lyapunov optimization to transform the stochastic optimization problem into a deterministic optimization problem, and proposed an online optimization algorithm that could dynamically adapt to the changes of the MEC system. By transforming the optimization problem into multiple subproblems, the optimal dynamic offloading and resource scheduling strategy could be obtained. Chen et al. [36] studied the energy saving scheduling and management of large-scale service computing systems. Due to the high dynamics and unpredictability of service request arrivals, the authors jointly considered the conflicts of performance, energy consumption and queue congestion in the article. Then, a distributed online scheduling algorithm was proposed. The algorithm does not require any prior information to achieve the trade-off between performance and energy efficiency.

Huang et al. [37] studied the energy efficiency of web service systems, and optimized web services by maximizing service quality and minimizing energy consumption costs. Through quantitative analysis of performance and energy consumption, they proposed a stochastic model of web service systems. The service problem was described as a Markov decision process problem, The related algorithms for solving large-scale web services was designed by using multi-agent technology, and the effectiveness of the algorithms was verified. Zhang et al. [38] studied the cost optimization problem of task scheduling in edge computing systems. The key point of the research was how to complete task offloading while minimizing the system cost. Therefore, the task scheduling problem was modeled as an optimization problem and proved to be an optimization problem. For the NP-hard problem, a two-stage task scheduling cost optimization algorithm was proposed. This algorithm could not only meet the QoS requirements of all tasks, but also minimize the system calculation overhead.

Li et al. [39] studied the performance and energy consumption of the IoT system, formulated the IoT service and system dynamic process by using the generalized queuing model. They used the Markov decision process for resource management and task scheduling and used the ordinal optimization technology to address the challenges brought by the large-scale IoT system. Finally, they realized the trade-off between energy consumption cost and QoS. Guo et al. [40] studied the problem of data partition calculation and offloading in MEC system consisting of an edge server and multiple users. Each mobile device deployed an energy collection module. It could continuously provide services for mobile device task offloading. Then, they established a model that joint local computing, task offloading, remote computing and energy collection. Because the energy causal constraint in the model would be coupled with task offloading, thus a task unloading algorithm based on Lyapunov optimization was proposed. By constructing the minimum drift penalty function, the energy causal constraint was eliminated, the stability of the energy queue of mobile devices was realized, the task can be offloaded continuously, and the balance between energy consumption and time delay was realized.

Zhang et al. [41] considered a UAV-assisted MEC system with stochastic computing tasks, and formulated the UAV computation offloading, resource allocation and trajectory scheduling problems as the average weighted energy minimization problem. In order to solve this joint optimization problem, Lyapunov optimization method was introduced. The optimization problem was decomposed into three subproblems and solved separately, and a joint optimization algorithm was proposed to obtain the minimum energy consumption by iterative solution.

Lai et al. [42] studied the stochastic marginal user allocation problem of the MEC system. Different from other studies, they looked the user allocation problem from the perspective of the service provider. Since the MEC environment and the user's arrival and departure were both random, the problem was modeled as a stochastic optimization problem to achieve a balance between system throughput and system cost. Therefore, an online optimization algorithm based on Lyapunov was proposed. Depending on whether the edge server had sufficient computing resources, part of the tasks could be allocated to the remote cloud to reduce queue delay time and lighten the burden of users.

Li et al. [43] studied the mobile cloud offloading scenario in heterogeneous wireless networks. Due to a large number of computation-intensive applications offloading tasks to the mobile cloud for computing, the computing pressure and computing overhead of the cloud computing center become larger. Therefore, the author modeled the user enrollment and offloading problem in heterogeneous wireless networks as a queue model. Finally, a multi-stage stochastic programming method was proposed to determine the optimal offloading workload. Bonadio et al. [44] proposed a Markov model with request revocation to design the computing power of SaaS server. They completed an evaluation of the MEC system that provided computing power to users in a limited area, and realized the allocation of the minimum number of processors. The reliability of the proposed model is verified by analysis, prediction and simulation under actual operating conditions.

Zhang et al. [45] discussed the impact of offloading strategy on the performance of IoT devices and established a multi queue model. The IoT devices was modeled as a queue system with two servers, which were the processing unit of local computation of IoT devices and the transmission module connected to edge servers. Then, they proposed two basic offloading strategies, and obtained the analytical solutions of the average task response time and energy consumption of the system. Duan et al. [46] combined with traditional cloud computing and MEC, realized the task offloading of UAV system. They optimized task scheduling and resource allocation under this heterogeneous cloud architecture, obtained the optimal solution by using queue theory and Lyapunov optimization. The minimum energy consumption under system stability constraints was realized.

Liu et al. [47] believed that it was very important to obtain the freshness of state update in time. They studied how to minimize the time to obtain state update, and used Lyapunov optimization technology to solve the time coupling problem between task generation and computation offloading. Therefore, they used virtual queue to obtain the optimal state update time. Fantacci et al. [48] studied the performance of cloud edge computing infrastructure and proposed a Markov queue system model. The model defaulted that the calculation request leaves in advance due to the arrival of the deadline. Then, they proposed a resource allocation method to maximize social welfare indicators.

Liu et al. [47] studied the stochastic control problem of task generation. That is, task generation, computation offloading and resource allocation were jointly optimized to minimize state update time under average energy constraints. They then used Lyapunov optimization technology to solve computation offloading and task generation. Then, an online task unloading algorithm was proposed to solve this coupling problem. The results of simulation experiments showed that the optimal performance could be achieved within the polynomial time complexity. Lin et al. [49] studied the optimization problem of computation offloading and resource allocation in edge computing networks. The goal was to minimize the average delay, and used the Lyapunov optimization method to transform the original problem into an upper bound optimization problem. Experiments showed that the distributed algorithm based on branch-and-bound method could achieve the balance between delay and queue.

Wu et al. [50] studied the task offloading problem in the blockchain scenario, and used edge computing and cloud computing to realize task offloading. Among them, MEC could provide low latency computing services, and MCC could provide high computing capacity. Combined with Lyapunov optimization algorithm, they proposed a dynamic task offloading algorithm, which could reduce energy consumption under the condition of low delay. Zhang et al. [51] proposed an online optimal auction to achieve optimal offloading of edge computing systems. In order to deal with the randomness of the energy harvesting process, Lyapunov optimization technology was introduced, and simulation experiments were used to verify the optimality of task allocation.

The above work has done a lot of research on task offloading of multi-user systems, but ignores the influence of channel conditions on the energy consumption of IoT devices, and this factor will have a great impact on the efficiency of task offloading. Based on the above-mentioned problems, we proposed a stochastic energy saving optimization framework to solve the task offloading problem of the multi-user MEC system. Under our stochastic energy saving optimization framework, there is no need to know the task arrival information and channel state information of IoT devices in advance, and an effective algorithm is designed to solve this optimization problem.

2.5 Summary

This chapter studies the efficient dynamic offloading problem of MEC system, and uses queue model and Lyapunov optimization technology to minimize offloading energy consumption to ensure the low delay requirements of IoT devices. In our system model, using Lyapunov optimization technology can make us not need to know the prior information of task offloading, which makes the optimization problem easier to understand. An online algorithm with polynomial time complexity is also proposed. It does not require a priori information about task arrival and channel conditions. EEDOA algorithm can realize the tradeoff between energy consumption and queue backlog by changing the value of parameter V. Experiments show that under the condition of ensuring the task queue balance of IoT devices, EEDOA algorithm can obtain an offloading strategy close to the optimal transmission energy consumption. Comparative experiments show that the algorithm can maintain the task queue length of IoT devices at a low level, and optimize the overall offloading energy consumption of the system.

References

1. D. Chen et al., S2M: A lightweight acoustic fingerprints-based wireless device authentication protocol. IEEE Internet Things J. **4**(1), 88–100, Feb. 2017. https://doi.org/10.1109/JIOT.2016.2619679
2. H. Duan, Y. Zheng, C. Wang, X. Yuan, Treasure collection on foggy islands: building secure network archives for Internet of Things. IEEE Internet Things J. **6**(2), 2637–2650 (2019). https://doi.org/10.1109/JIOT.2018.2872461
3. L.F. Bittencourt, J. Diaz-Montes, R. Buyya, O.F. Rana, M. Parashar, Mobility-aware application scheduling in fog computing. IEEE Cloud Comput. **4**(2), 26–35 (2017). https://doi.org/10.1109/MCC.2017.27
4. W. Shi, J. Cao, Q. Zhang, Y. Li, L. Xu, Edge computing: vision and challenges. IEEE Internet Things J. **3**(5), 637–646 (2016). https://doi.org/10.1109/JIOT.2016.2579198
5. X. Chen, Q. Shi, L. Yang, J. Xu, ThriftyEdge: resource-efficient edge computing for intelligent IoT applications. IEEE Netw. **32**(1), 61–65 (2018). https://doi.org/10.1109/MNET.2018.1700145

6. A. Kapsalis, P. Kasnesis, I.S. Venieris, D.I. Kaklamani, C.Z. Patrikakis, A cooperative fog approach for effective workload balancing. IEEE Cloud Comput. **4**(2), 36–45 (2017). https://doi.org/10.1109/MCC.2017.25
7. H. Wu, Y. Sun, K. Wolter, Energy-efficient decision making for mobile cloud offloading. IEEE Trans. Cloud Comput. **8**(2), 570–584 (2020). https://doi.org/10.1109/TCC.2018.2789446
8. Y. Wu, J. Chen, L.P. Qian, J. Huang, X.S. Shen, Energy-aware cooperative traffic offloading via device-to-device cooperations: an analytical approach. IEEE Trans. Mob. Comput. **16**(1), 97–114 (2017). https://doi.org/10.1109/TMC.2016.2539950
9. F. Liu, P. Shu, J.C.S. Lui, AppATP: an energy conserving adaptive mobile-cloud transmission protocol. IEEE Trans. Comput. **64**(11), 3051–3063 (2015). https://doi.org/10.1109/TC.2015.2401032
10. O. Muñoz, A. Pascual-Iserte, J. Vidal, Optimization of radio and computational resources for energy efficiency in latency-constrained application offloading. IEEE Trans. Veh. Technol. **64**(10), 4738–4755 (2015). https://doi.org/10.1109/TVT.2014.2372852
11. M.R. Palattella et al., Internet of Things in the 5G era: enablers, architecture, and business models. IEEE J. Sel. Areas Commun. **34**(3), 510–527 (2016). https://doi.org/10.1109/JSAC.2016.2525418
12. X. Lyu et al., Optimal schedule of mobile edge computing for internet of things using partial information. IEEE J. Sel. Areas Commun. **35**(11), 2606–2615 (2017). https://doi.org/10.1109/JSAC.2017.2760186
13. Y. Mao, J. Zhang, S.H. Song, K.B. Letaief, Stochastic joint radio and computational resource management for multi-user mobile-edge computing systems. IEEE Trans. Wirel. Commun. **16**(9), 5994–6009 (2017). https://doi.org/10.1109/TWC.2017.2717986
14. K. Wang, K. Yang, C.S. Magurawalage, Joint energy minimization and resource allocation in C-RAN with mobile cloud. IEEE Trans. Cloud Comput. **6**(3), 760–770 (2018). https://doi.org/10.1109/TCC.2016.2522439
15. X. Chen, L. Jiao, W. Li, X. Fu, Efficient multi-user computation offloading for mobile-edge cloud computing. IEEE/ACM Trans. Netw. **24**(5), 2795–2808 (2016). https://doi.org/10.1109/TNET.2015.2487344
16. Y. Yu, J. Zhang, K.B. Letaief, Joint subcarrier and CPU time allocation for mobile edge computing, in *2016 IEEE Global Communications Conference (GLOBECOM)* (2016), pp. 1–6. https://doi.org/10.1109/GLOCOM.2016.7841937
17. X. Lyu, H. Tian, C. Sengul, P. Zhang, Multiuser joint task offloading and resource optimization in proximate clouds. IEEE Trans. Veh. Technol. **66**(4), 3435–3447 (2017). https://doi.org/10.1109/TVT.2016.2593486
18. C. You, K. Huang, H. Chae, Energy efficient mobile cloud computing powered by wireless energy transfer. IEEE J. Sel. Areas Commun. **34**(5), 1757–1771 (2016). https://doi.org/10.1109/JSAC.2016.2545382
19. Y. Chen, Z. Liu, Y. Zhang, Y. Wu, X. Chen, L. Zhao, Deep reinforcement learning-based dynamic resource management for mobile edge computing in industrial Internet of Things. IEEE Trans. Ind. Inform. **17**(7), 4925–4934 (2021). https://doi.org/10.1109/TII.2020.3028963
20. Y. Zhang, X. Chen, Y. Chen, Z. Li, J. Huang, Cost efficient scheduling for delay-sensitive tasks in edge computing system, in *2018 IEEE International Conference on Services Computing (SCC)* (2018), pp. 73–80. https://doi.org/10.1109/SCC.2018.00017
21. X. Lyu et al., Distributed online optimization of fog computing for selfish devices with out-of-date information. IEEE Trans. Wirel. Commun. **17**(11), 7704–7717 (2018). https://doi.org/10.1109/TWC.2018.2869764
22. J. Kwak, Y. Kim, J. Lee, S. Chong, DREAM: dynamic resource and task allocation for energy minimization in mobile cloud systems. IEEE J. Sel. Areas Commun. **33**(12), 2510–2523 (2015). https://doi.org/10.1109/JSAC.2015.2478718
23. Z. Jiang, S. Mao, Energy delay tradeoff in cloud offloading for multi-core mobile devices. IEEE Access **3**, 2306–2316 (2015). https://doi.org/10.1109/ACCESS.2015.2499300

24. Y. Chen, Y. Zhang, Y. Wu, L. Qi, X. Chen, X. Shen, Joint task scheduling and energy management for heterogeneous mobile edge computing with hybrid energy supply. IEEE Internet Things J. **7**(9), 8419–8429 (2020). https://doi.org/10.1109/JIOT.2020.2992522
25. Y. Chen, C. Lin, J. Huang, X. Xiang, X. Shen, Energy efficient scheduling and management for large-scale services computing systems. IEEE Trans. Serv. Comput. **10**(2), 217–230 (2017). https://doi.org/10.1109/TSC.2015.2444845
26. X. Chen, Y. Zhang, Y. Chen, Cost-efficient request scheduling and resource provisioning in multiclouds for Internet of Things. IEEE Internet Things J. **7**(3), 1594–1602 (2020). https://doi.org/10.1109/JIOT.2019.2948432
27. Y. Chen, N. Zhang, Y. Zhang, X. Chen, Dynamic computation offloading in edge computing for Internet of Things. IEEE Internet Things J. **6**(3), 4242–4251 (2019). https://doi.org/10.1109/JIOT.2018.2875715
28. Y. Chen, J. Huang, X. Xiang, C. Lin, Energy efficient dynamic service selection for large-scale web service systems, in *2014 IEEE International Conference on Web Services* (2014), pp. 558–565. https://doi.org/10.1109/ICWS.2014.84
29. Y. Chen, J. Huang, C. Lin, J. Hu, A partial selection methodology for efficient QoS-aware service composition. IEEE Trans. Serv. Comput. **8**(3), 384–397 (2015). https://doi.org/10.1109/TSC.2014.2381493
30. G. Chen, J. Huang, B. Cheng, J. Chen, A social network based approach for IoT device management and service composition, in *2015 IEEE World Congress on Services* (2015), pp. 1–8. https://doi.org/10.1109/SERVICES.2015.9
31. X. Chen, Z. Liu, Y. Chen, Z. Li, Mobile edge computing based task offloading and resource allocation in 5G ultra-dense networks. IEEE Access **7**, 184172–184182 (2019). https://doi.org/10.1109/ACCESS.2019.2960547
32. S. Chen, X. Chen, Y. Chen, Z. Li, Distributed computation offloading based on stochastic game in multi-server mobile edge computing networks, in *2019 IEEE International Conference on Smart Internet of Things (SmartIoT)* (2019), pp. 77–84. https://doi.org/10.1109/SmartIoT.2019.00021
33. P. Yao, X. Chen, Y. Chen, Z. Li, Deep reinforcement learning based offloading scheme for mobile edge computing, in *2019 IEEE International Conference on Smart Internet of Things (SmartIoT)* (2019), pp. 417–421. https://doi.org/10.1109/SmartIoT.2019.00074
34. Y. Yang, X. Chen, Y. Chen, Z. Li, Green-oriented offloading and resource allocation by reinforcement learning in MEC, in *2019 IEEE International Conference on Smart Internet of Things (SmartIoT)* (2019), pp. 378–382. https://doi.org/10.1109/SmartIoT.2019.00066
35. F. Zhao, Y. Chen, Y. Zhang, Z. Liu, X. Chen, Dynamic offloading and resource scheduling for mobile-edge computing with energy harvesting devices. IEEE Trans. Netw. Serv. Manag. **18**(2), 2154–2165 (2021). https://doi.org/10.1109/TNSM.2021.3069993
36. Y. Chen, C. Lin, J. Huang, X. Xiang, X. Shen, Energy efficient scheduling and management for large-scale services computing systems. IEEE Trans. Serv. Comput. **10**(2), 217–230 (2017). https://doi.org/10.1109/TSC.2015.2444845
37. J. Huang, C. Lin, Agent-based green web service selection and dynamic speed scaling, in *2013 IEEE 20th International Conference on Web Services* (2013), pp. 91–98. https://doi.org/10.1109/ICWS.2013.22
38. Y. Zhang, X. Chen, Y. Chen, Z. Li, J. Huang, Cost efficient scheduling for delay-sensitive tasks in edge computing system, in *2018 IEEE International Conference on Services Computing (SCC)* (2018), pp. 73–80. https://doi.org/10.1109/SCC.2018.00017
39. S. Li, J. Huang, Energy efficient resource management and task scheduling for IoT services in edge computing paradigm, in *2017 IEEE International Symposium on Parallel and Distributed Processing with Applications and 2017 IEEE International Conference on Ubiquitous Computing and Communications (ISPA/IUCC)* (2017), pp. 846–851. https://doi.org/10.1109/ISPA/IUCC.2017.00129
40. M. Guo, W. Wang, X. Huang, Y. Chen, L. Zhang, L. Chen, Lyapunov-based partial computation offloading for multiple mobile devices enabled by harvested energy in MEC. IEEE Internet Things J.. https://doi.org/10.1109/JIOT.2021.3118016

41. J. Zhang et al., Stochastic computation offloading and trajectory scheduling for UAV-assisted mobile edge computing. IEEE Internet Things J. **6**(2), 3688–3699 (2019). https://doi.org/10.1109/JIOT.2018.2890133
42. P. Lai et al., Dynamic user allocation in stochastic mobile edge computing systems. IEEE Trans. Serv. Comput. (2021). https://doi.org/10.1109/TSC.2021.3063148
43. Y. Li, S. Xia, M. Zheng, B. Cao, Q. Liu, Lyapunov optimization based trade-off policy for mobile cloud offloading in heterogeneous wireless networks. IEEE Trans. Cloud Comput. (2019). https://doi.org/10.1109/TCC.2019.2938504
44. A. Bonadio, F. Chiti, R. Fantacci, Performance analysis of an edge computing SaaS system for mobile users. IEEE Trans. Veh. Technol. **69**(2), 2049–2057 (2020). https://doi.org/10.1109/TVT.2019.2957938
45. Y. Zhang, B. Feng, W. Quan, G. Li, H. Zhou, H. Zhang, Theoretical analysis on edge computation offloading policies for IoT devices. IEEE Internet Things J. **6**(3), 4228–4241 (2019). https://doi.org/10.1109/JIOT.2018.2875599
46. R. Duan, J. Wang, J. Du, C. Jiang, T. Bai, Y. Ren, Power-delay trade-off for heterogenous cloud enabled multi-UAV systems, in *ICC 2019 - 2019 IEEE International Conference on Communications (ICC)* (2019), pp. 1–6. https://doi.org/10.1109/ICC.2019.8761377
47. L. Liu, X. Qin, Z. Zhang, P. Zhang, Joint task offloading and resource allocation for obtaining fresh status updates in multi-device MEC systems. IEEE Access **8**, 38248–38261 (2020). https://doi.org/10.1109/ACCESS.2020.2976048
48. R. Fantacci, B. Picano, Performance analysis of a delay constrained data offloading scheme in an integrated cloud-fog-edge computing system. IEEE Trans. Veh. Technol. **69**(10), 12004–12014 (2020). https://doi.org/10.1109/TVT.2020.3008926
49. R. Lin et al., Distributed optimization for computation offloading in edge computing. IEEE Trans. Wirel. Commun. **19**(12), 8179–8194 (2020). https://doi.org/10.1109/TWC.2020.3019805
50. H. Wu, K. Wolter, P. Jiao, Y. Deng, Y. Zhao, M. Xu, EEDTO: an energy-efficient dynamic task offloading algorithm for blockchain-enabled IoT-edge-cloud orchestrated computing. IEEE Internet Things J. **8**(4), 2163–2176 (2021). https://doi.org/10.1109/JIOT.2020.3033521
51. D. Zhang et al., Near-optimal and truthful online auction for computation offloading in green edge-computing systems. IEEE Trans. Mob. Comput. **19**(4), 880–893 (2020). https://doi.org/10.1109/TMC.2019.2901474

Chapter 3
Energy Efficient Offloading and Frequency Scaling for Internet of Things Devices

3.1 System Model and Problem Formulation

With the rapid development of mobile applications, mobile services based on the Internet of Things (IoT) are becoming more diverse . Due to the diversity of service requirements, mobile applications become more complex. While this change has brought more convenience, it has also led to more intensive computing [1–3]. Therefore, the requirements for mobile devices are getting higher and higher. However, due to the limited computing power and battery capacity of mobile devices, it is not practical to fully handle the tasks generated by these applications locally. In this case, mobile cloud computing (MCC) came into being. MCC can receive computing tasks from mobile devices and process them on its own cloud resources [4–6]. By offloading computing tasks to the MCC for processing, local computing is relieved. However, as a cloud server, it is typically located away from IoT devices. As a result, offloading tasks to the cloud consumes a lot of energy and incurs long latency [7]. To solve these problems, mobile edge computing (MEC) is thus developed [8, 9]. The basic idea behind MEC is to deploy servers at the edge of IoT devices [10]. As the distance between the server and the mobile device is greatly shortened, not only the transmission delay is reduced, but also the energy consumption is less. In this computing mode, the energy consumption of IoT devices will be reduced and the quality of user experience will be improved [11, 12].

In an IoT environment, mobile devices may need to handle many tasks. Moreover, most of these tasks are computation-intensive. Thus, the computation to complete these tasks needs to consume a lot of computing resources. However, IoT devices have limited computing resources. As a result, these computation-intensive tasks can drain mobile device's battery and reduce battery life [13–15]. When too many computation-intensive tasks are processed locally, a large amount of power consumption will be generated and the battery life will be affected. In order to prolong battery life, academia and industry began to pay continuous attention to the relationship between computation offloading and CPU-cycle frequency conversion

Y. Chen et al., *Energy Efficient Computation Offloading in Mobile Edge Computing*, Wireless Networks, https://doi.org/10.1007/978-3-031-16822-2_3

[11, 16, 17]. The decision to calculate the amount of offloading will have a great impact on the energy consumption of the system. At the same time, the conditions of wireless communication channel and tail energy also affect the offloading decision [13, 18]. When wireless channel conditions are good, mobile devices can offload tasks in batches, which is of great help to reduce transmission energy consumption. This is not the only way to reduce transmission energy consumption. Another way is to dynamically scale the voltage frequency [19]. When computing tasks are processed locally, computing power consumption is mainly related to the CPU cycle frequency of mobile devices. That is, when the CPU cycle frequency of mobile devices is higher, the CPU power is higher, and more energy is consumed in the same amount of time. Therefore, the local CPU cycle frequency can be appropriately reduced. In this way, not only the task calculation will not be affected, but also reduce the power consumption of local computing. Although this method can save energy, it will lead to a large queuing delay of computing tasks. When this delay is too great, it can make IoT devices unstable. Therefore, in order to achieve the balance between power consumption and performance, it is necessary to determine the optimal task allocation and CPU frequency decision.

However, due to the task generation of each application on mobile devices is uncertain and dynamic. Therefore, it is very difficult to design such a combination of task allocation and frequency scaling algorithm. First, we cannot predict the arrival information of the task. Thus, we cannot determine how much computing resources should be allocated to the local CPU. No matter how many computing resources are allocated to local computing, queuing delays are likely to occur [20]. Second, the current queue state and wireless channel conditions will affect the task allocation decision [21]. When the queue status is unstable, the transfer of tasks may not be continuous, and the idle time in the middle will increase the system energy consumption. When the channel is in poor condition, it may take more time to transmit data, resulting in greater power consumption. However, the arrival situation of the task and the conditions of the wireless channel are affected by many aspects, including the characteristics of the MEC system itself and the influence of the external environment [22]. Because these prior information is random, it is very difficult to obtain it accurately in advance. Therefore, it is very difficult to design an algorithm that can not only adapt to the arrival of the task but also not be affected by the dynamic change of the channel.

The above problem occurs when a task is offloaded to the MEC for calculation. In order to solve the above problems, we jointly study task allocation and CPU cycle allocation in MEC system. Because the task queue length cannot grow indefinitely, we can seek an upper bound on queue delay. After obtaining the upper bound of queuing delay, we can obtain the minimum average energy consumption of MEC system. We propose a stochastic optimization problem to optimize the energy consumption of the system and reduce the energy consumption of mobile devices. Then, we transform the joint optimization problem into deterministic optimization problem by using stochastic optimization technique, and decompose it into two deterministic subproblems. Then, we design the COFSEE algorithm to solve these sub-problems. COFSEE algorithm does not require prior knowledge of mission arrival and wireless channel information. Then, we use mathematical analysis to prove the effectiveness of COFSEE. Finally, the performance of COFSEE is evaluated by experiments.

3.1.1 Network Model

In our model, a system consisting of a mobile device and a base station is considered. Among them, N different applications run simultaneously on the mobile device, and the application set is represented as $N = \{1, 2, \cdots, n\}$. An edge server is deployed on the base station to provide computing services to mobile devices. The mobile device can offload tasks to the MEC server for processing via the wireless channel. Since the MEC server has powerful computing power, it can share more computing tasks for mobile devices. As a result, the pressure on the mobile device to handle tasks is reduced and battery life is extended. At the same time, we consider a discrete time slot system, i.e., $t = \{0, 1 \cdots, T - 1\}$. We set the length of each time slot as τ. In each time slot t, mobile device makes a task assignment decision. Before making a decision, the mobile device first observes the wireless channel status and looks at the backlog of the queue. Based on these two points, allocation decisions are then jointly made (determining the amount of local computation and offloading computation), and the local CPU cycle frequency is determined. The main symbols and meanings in the system model are given in Table 3.1.

3.1.2 Task Model

In this chapter, the tasks generated by N applications are computation-intensive. We divide the generated task into several parts to execute. Among them, some tasks are performed locally by the local device, and some tasks are performed by offloading to the MEC. When the amount of processing tasks (in bits) is larger, the processing time required is also larger. Therefore, when the amount of tasks that arrive is larger, the complexity of task processing is also larger [8, 11, 17].

The size of the task amount generated by the i-th application in time slot t is set to $A_i(t)$. Since each application is different and they do not interfere with each other, the task generation amount $A_i(t)$ of each application is also different. We set the amount of tasks performed locally by the i-th application in time slot t as $D_i^l(t)$. And set the number of CPU cycles required to calculate 1 bit data as φ_i. The value of φ_i can be obtained by calling the graph analysis method. The value of φ_i is determined by the nature of the application. Since each application is different from each other, the φ_i values of n applications are also different. Therefore, the sum of the CPU cycles required by n applications in time slot t is $\sum_{i=1}^{n} \varphi_i D_i^l(t)$. The CPU cycle frequency of the mobile device is set to $f(t)$, and the maximum value of the CPU frequency is f^{max}, which is expressed as,

Table 3.1 Notations and definitions

Notion	Definition
$\mathbf{N}$	Applications set
τ	Time slot length
P	The mobile device's transmit power
$h(t)$	The wireless channel's power gain in slot t
B	Bandwidth of the wireless channel
σ^2	Noise power at the receiver
$R_i(t)$	Task offloading rate in slot t
$A_i(t)$	Amount of input tasks from the i-th application in slot t
$D_i^l(t)$	Amount of tasks processed locally from the i-th application in slot t
$D_i^r(t)$	Amount of offloaded tasks from the i-th application in slot t
$D_i(t)$	Amount of processed tasks from the i-th application in slot t
φ_i	The required CPU cycles to compute 1 bit i-th application task
$f(t)$	CPU-cycle frequency in slot t
$E^l(t)$	Energy consumption for local execution in slot t
$E^r(t)$	Energy consumption for transmission in slot t
$E^a(t)$	Tail energy consumption in slot t
$E_{total}(t)$	Total energy consumption in slot t
$Q_i(t)$	Queue backlog of the application i in slot t

$$f(t) \leq f^{max}. \tag{3.1}$$

Due to the limited processing power of mobile devices, the amount of tasks used for local processing cannot exceed the local maximum processing capacity [23]. The amount of local computation needs to be satisfied,

$$\sum_{i=1}^{n} \varphi_i D_i^l(t) \leq f(t)\tau. \tag{3.2}$$

The amount of tasks that the i-th application offloads to the MEC in time slot t is denoted as $D_i^r(t)$. Then the total offloading task amount of the mobile device in the time slot t is $\sum_{i=1}^{n} D_i^r(t)$. The achievable data transfer rate when application offloading tasks to MEC can be obtained by,

$$R(t) = B\log_2(1 + \frac{Ph(t)}{\sigma^2}). \tag{3.3}$$

In (3.3), P represents the transmit power of the mobile device, $h(t)$ is the wireless channel power gain, B represents the wireless channel bandwidth, and σ^2 represents the channel noise power.

Since the transmission power of the device is limited, the size of the device offloading task is also limited. In the time slot t, the offloadable computation of the mobile device is expressed as follows,

$$\sum_{i=1}^{n} D_i^r(t) \leq R(t)\tau. \tag{3.4}$$

Looking back at the above, we have given the representation of the amount of local computation and offloaded computation. Therefore, we can get the total task computation of the i-th application in time slot t, and satisfy,

$$D_i(t) = D_i^l(t) + D_i^r(t). \tag{3.5}$$

3.1.3 Queuing Model

The task queue backlog of the i-th application is denoted as $Q_i(t)$. Therefore, the evolution of the queue backlog for the i-th application is denoted as,

$$Q_i(t+1) = [Q_i(t) - D_i(t), 0]^+ + A_i(t). \tag{3.6}$$

It should be noted that when the application's queue backlog is larger, its task delay will also be larger. Therefore, in order to achieve the best performance for each application, it is necessary to ensure that the queue backlog of the application is stable. That is, it needs to meet,

$$q_i = \lim_{T\to\infty} \frac{1}{T} \sum_{t=0}^{T-1} \mathbf{E}\{Q_i(t)\} < \varepsilon, \exists\, \varepsilon \in \mathbb{R}^+, \tag{3.7}$$

where q_i represents the average queue backlog for all time slots. Here, we ignore the overhead of partitioning, migration and returning results, just as most literatures on computation offloading.

3.1.4 Energy Consumption Model

In this part, we will give a representation of system energy consumption. When the mobile device processes the tasks generated by the application, part of the tasks is used for local computing and part of the tasks is used for transmission offload. Therefore, device energy consumption will include local computing energy consumption and transmission offload energy consumption. However, after the device completes the data transfer, it will run for a period of time (tail time). During the tail time, the device will remain in a high power state, thus generating tail energy. In 3G networks, tail time is introduced to reduce high signaling overhead [18]. At the same time, tail time was also introduced in 4G LTE [24].

For local computing, its energy consumption is closely related to the mobile device's CPU [25]. At the same time, it is also related to the amount of local computation. Therefore, the local computing energy consumption is jointly determined by the CPU cycle frequency of the mobile device and the local computing amount [26]. We define the local computing energy consumption as $E^l(t)$, which is expressed as follows,

$$E^l(t) = \xi f^2(t) \sum_{i=1}^{n} \varphi_i D_i^l(t), \tag{3.8}$$

where ξ is the effective switched capacitor, which is determined by the chip structure [27].

The energy consumed to transmit offloaded data is related to the transmission power and the amount of offload computation. We use the product of transmission power and transmission time to express the transmission offload energy consumption. We define the data transmission energy consumption as $E^r(t)$, which is expressed as follows,

$$E^r(t) = P \sum_{i=1}^{n} D_i^r(t)/R(t). \tag{3.9}$$

Radio Resource Control (RRC) protocol is an important way of channel allocation in current cellular networks [28, 29]. Specifically, the RRC protocol consists of three states, including: data transmission (DT), tail (TA) and idle (ID) [24, 30]. The power consumption of these three states is denoted as P, P_T, P_I, respectively. The energy consumption of tasks in a cellular network can be modeled as follows. When the transmitted data arrives, if the radio power is in the TA or ID phase, it will wake up the higher power DT. When the incoming data has been transmitted, the radio power is switched to the TA again. After that, if there is no data to transmit, the power of the radio will remain at TA for a period of time, the size of this period is denoted as δ_T, and then switch to ID.

The time to stay after completing the last data transfer is set to as Δt. Therefore, if there are still computing tasks to be transmitted in the time slot t, the tail time of this time slot is 0. Otherwise, it means that no task transmission is required, and it can be known that the tail time is τ. Therefore, the tail time in slot t is calculated as follows,

$$\Delta t' = \begin{cases} 0, & \sum_{i=1}^{n} D_i^r(t) > 0 \\ \tau, & otherwise. \end{cases} \tag{3.10}$$

Then, define $E^a(t)$ as the tail energy consumption in time slot t, which can be derived by (3.11).

$$E^a(t) = \begin{cases} 0, & \Delta t \geq \delta_T \ or \ \Delta t' = 0 \\ P_T * \tau, & \Delta t < \delta_T, \Delta t + \Delta t' \leq \delta_T \\ P_T * (\delta_T - \Delta t), & otherwise. \end{cases} \tag{3.11}$$

So far, we can get the energy consumption of the device in the time slot t, which is expressed as follows,

$$E_{total}(t) = \xi f^2(t) \sum_{i=1}^{n} \varphi_i D_i^l(t) + P \sum_{i=1}^{n} D_i^r(t)/R(t) + E^a(t). \tag{3.12}$$

Since the wireless channel state is constantly changing, even the same task may consume different energy when transmitting at different times. Typically, the tail energy consumption of a device is related to the computation offloading decision for each time slot. Therefore, we focus on the long-term average energy consumption of the equipment, as follows,

$$\bar{E} = \lim_{T \to \infty} \frac{1}{T} \sum_{t=0}^{T-1} \mathbf{E}\{E_{total}(t)\}. \tag{3.13}$$

3.1.5 Problem Formulation

In this part, our goal is to optimize the average energy consumption of mobile devices. The energy consumption of mobile devices consists of local computing energy consumption and data transmission energy consumption. On the one hand, local computing energy consumption can be reduced by reducing the amount of local computation and reducing the local CPU cycle frequency. On the other hand, the quality of the channel conditions will also have a great impact on the energy consumption of data transmission. When the channel conditions are good, it will take less time to transmit data, and thus less energy will be consumed [13]. In addition, reducing the amount of computation for offloading can also reduce energy consumption during data transmission. Finally, tail energy consumption can be reduced by transmitting data in batches [18]. Although the above methods can reduce energy consumption to a certain extent, they will increase the queue backlog of the application and make the device unstable. Therefore, to ensure a balance between energy consumption and queue backlog, the application's task delay needs to be controlled. Our goal is to minimize the average energy consumption. The optimization problem is,

$$
\begin{aligned}
\min_{\mathbf{D}^l(t),\mathbf{D}^r(t),f(t)} \quad & \bar{E} = \lim_{T\to\infty} \frac{1}{T} \sum_{t=0}^{T-1} \mathbf{E}\{E_{total}(t)\}. \\
s.t. \quad & f(t) \leq f^{max}. \\
& \sum_{i=1}^{n} \varphi_i D_i^l(t) \leq f(t)\tau. \\
& \sum_{i=1}^{n} D_i^r(t) \leq R(t)\tau. \\
& q_i = \lim_{T\to\infty} \frac{1}{T} \sum_{t=0}^{T-1} \mathbf{E}\{Q_i(t)\} < \varepsilon, \exists\, \varepsilon \in \mathbb{R}^+.
\end{aligned} \tag{3.14}
$$

Due to the dynamic nature of task arrivals, we can neither determine when tasks arrive nor the amount of tasks that arrive. At the same time, the wireless channel is affected by various factors, so that the state of the channel cannot be acquired in time. Therefore, optimization problem (3.14) is a stochastic optimization problem.

3.2 COFSEE: Computation Offloading and Frequency Scaling for Energy Efficiency of Internet of Things Devices

In this part, we design a computation offloading and frequency scaling algorithm based on the energy efficiency of IoT devices. The algorithm can simultaneously minimize energy consumption and dynamically scale the CPU frequency. However, our proposed optimization problem (3.14) is a stochastic optimization problem. Therefore, it is difficult for us to solve this optimization problem. To this end, we introduce the Lyapunov optimization technique [31, 32]. With this technique, we decouple problem (3.14) into two deterministic optimization subproblems. Since the task generation of the application is stochastic, and the wireless channel state is constantly changing over time, it is almost impossible to know exactly the task generation and channel state information at each moment. At the same time, in actual scenarios, such information is also difficult to obtain. Our COFSEE algorithm takes these factors into account, so it can better solve this task offloading problem. Not only that, we also consider the effectiveness of the algorithm, the COFSEE algorithm has a lower time complexity.

3.2.1 Problem Transformation

To ensure a balance between energy consumption and queue backlog, we use Lyapunov optimization techniques to transform the problem to be solved into an easy-to-solve deterministic optimization problem. We define the Lyapunov optimization function as follows,

$$L(\Theta(t)) = \frac{1}{2}\sum_{i=1}^{n} Q_i^2(t), \tag{3.15}$$

where $\Theta(t) = [Q_1(t), Q_2(t), \cdots, Q_n(t)]$ represents the queue backlog matrix for all applications.

In (3.15), according to the definition, if the task queue backlog of an application is too large, the total queue backlog of the device will also become large. Therefore, if want to keep the overall queue backlog for mobile devices low, need to ensure that each application maintains a small task queue backlog. Here, our goal is to reduce the total queue length. Therefore, we define the conditional Lyapunov drift function as follows,

$$\Delta(\Theta(t)) = \mathbf{E}\{L(\Theta(t+1)) - L(\Theta(t))|\Theta(t)\}. \tag{3.16}$$

To be able to analyze both energy consumption and mobile device performance, we define a drift plus penalty function as follows,

$$\Delta_V(\Theta(t)) = \Delta(\Theta(t)) + V\mathbf{E}\{E_{total}(t)|\Theta(t)\}, \tag{3.17}$$

where $V \geq 0$ is a control coefficient, which is a parameter used to measure energy consumption and queue stability. The larger the value of V, the greater the weight of energy consumption. In practical applications, since it may be more inclined to one of energy consumption and queue backlog, the value of V can be adjusted to achieve this. In order to minimize the drift plus penalty function, the upper bound of the drift plus penalty function is given in Theorem 3.1.

Theorem 3.1 *In each time slot t, for any V, if there exist the upper bounds of $A_i(t)$ and $R(t)$, which are A_i^{max} and R^{max}, the $\Delta_V(\Theta(t))$ would satisfy,*

$$\begin{aligned}
&\Delta(\Theta(t)) + V\mathbf{E}\{E_{total}(t)|\Theta(t)\} \\
&\leq C + \sum_{i=1}^{n} Q_i(t)\mathbf{E}\{A_i(t) - D_i(t)|\Theta(t)\} \\
&+ V\xi\mathbf{E}\{f^2(t)\sum_{i=1}^{n}\varphi_i D_i^l(t)|\Theta(t)\} \\
&+ VP\sum_{i=1}^{n}\mathbf{E}\{D_i^r(t)/R(t)|\Theta(t)\} + V\mathbf{E}\{E^a(t)|\Theta(t)\},
\end{aligned} \tag{3.18}$$

where $C = \frac{1}{2}\sum_{i=1}^{n}[(A_i^{max})^2 + (\frac{f^{max}\tau}{\varphi_i} + R^{max}\tau)^2]$ is a constant.

The following is the proof of Theorem 3.1.

Proof Using $(\max[a-b,0])^2 \leq a^2+b^2-2ab$ and Eq. (3.6) for any $a, b \geq 0$, we have,

$$Q_i^2(t+1) \leq Q_i^2(t) + D_i^2(t) + A_i(t)^2 - 2Q_i(t)D_i(t) + 2A_i(t)[Q_i(t) - D_i(t), 0]^+. \tag{3.19}$$

The actual amount of tasks that application i can process in time slot t is denoted $\bar{D}_i(t)$. Thus, we have,

$$\bar{D}_i(t) = \begin{cases} Q_i(t), & Q_i(t) \leq D_i(t) \\ D_i(t), & otherwise. \end{cases} \tag{3.20}$$

So we can get by (3.20) that $\max[Q_i(t)-D_i(t),0] = Q_i(t)-\bar{D}_i(t)$. Then (3.19) is rewritten as,

$$Q_i^2(t+1) \leq Q_i^2(t) + D_i^2(t) + A_i^2(t) + 2Q_i(t)[A_i(t) - D_i(t)] - 2\bar{D}_i(t)A_i(t). \tag{3.21}$$

Since $\bar{D}_i(t)$ and $A_i(t)$ are non-negative, it holds (3.22).

$$\frac{1}{2}[Q_i^2(t+1) - Q_i^2(t)] \leq \frac{1}{2}[A_i^2(t) + D_i^2(t)] + Q_i(t)[A_i(t) - D_i(t)]. \tag{3.22}$$

Summing over all the applications on (2.18) and taking conditional expectations, (3.23) holds.

$$\Delta(\Theta(t)) \leq \frac{1}{2}\sum_{i=1}^{n} \mathbf{E}\{A_i^2(t) + D_i^2(t)|\Theta(t)\} + \sum_{i=1}^{n} Q_i(t)\mathbf{E}\{A_i(t) - D_i(t)|\Theta(t)\}. \tag{3.23}$$

Since for any application i, there holds that $R_i(t) \leq R_i^{max}$ and $f(t) \leq f^{max}$. Thus, we can get that $D_i^l(t) \leq \frac{f^{max}\tau}{\varphi_i}$ and $D_i^r(t) \leq R^{max}\tau$. Since $D_i(t) = D_i^l(t) + D_i^r(t)$, therefore, we can get (3.24).

$$D_i(t) \leq \frac{f^{max}\tau}{\varphi_i} + R^{max}\tau. \tag{3.24}$$

Combining Eq. (3.24) and $A_i(t) \leq A_i^{max}$, we can get the following conclusion.

$$\sum_{i=1}^{n} \mathbf{E}\{A_i^2(t) + D_i^2(t)|\Theta(t)\} \leq \sum_{i=1}^{n}[(A_i^{max})^2 + (\frac{f^{max}\tau}{\varphi_i} + R^{max}\tau)^2]. \tag{3.25}$$

Let C equal to $\frac{1}{2}\sum_{i=1}^{n}[(A_i^{max})^2+(\frac{f^{max}\tau}{\varphi_i}+R^{max}\tau)^2]$, and add $V\mathbf{E}\{E_{total}(t)|\Theta(t)\}$ to (3.23). Then, substituting it into (3.12), we get (3.18).

3.2.2 *Optimal Frequency Scaling*

From the previous description, we already know that the frequency of the CPU will have a certain impact on the energy consumption of mobile devices. Therefore, in order to minimize energy consumption, we need to obtain optimal CPU frequency scaling. In the previous part, we have obtained the upper bound on COFSEE. Since both C and $A_i(t)$ can be considered as a constant, the rewrite minimization problem is as follows,

$$\begin{aligned}
\min_{\mathbf{D}^l(t),\mathbf{D}^r(t),f(t)} \quad & \{\sum_{i=1}^{n}[V\xi f^2(t)\varphi_i D_i^l(t) - Q_i(t)D_i^l(t)] \\
& + \sum_{i=1}^{n}[VP/R(t) - Q_i(t)]D_i^r(t) + VE^a(t)\}. \\
s.t. \quad & f(t) \leq f^{max}. \\
& \sum_{i=1}^{n}\varphi_i D_i^l(t) \leq f(t)\tau. \\
& \sum_{i=1}^{n} D_i^r(t) \leq R(t)\tau.
\end{aligned} \tag{3.26}$$

The rewritten minimization problem is still not easy to solve. Therefore, we first determine the optimal CPU cycle frequency. Lemma 1 below gives the optimal CPU cycle frequency representation.

Lemma 3.1 *When determining the amount of local computation $\sum_{i=1}^{n}\varphi_i D_i^l(t)$, the optimal CPU-cycle frequency $f^*(t)$ for Problem (3.26) can be obtained as follows,*

$$f^*(t) = \sum_{i=1}^{n}\varphi_i D_i^l(t)/\tau. \tag{3.27}$$

Proof When $\sum_{i=1}^{n}\varphi_i D_i^l(t)$ is determined, then Problem (3.26) is transformed into a non-decreasing function of $f(t)$. Therefore, to minimize Problem (3.26), the value of $f(t)$ needs to be minimized. At the same time, we know from Eq. (3.2) that $f(t) \geq \sum_{i=1}^{n} \varphi_i D_i^l(t)/\tau$. Therefore, the optimal CPU cycle frequency $f^*(t) = \sum_{i=1}^{n} \varphi_i D_i^l(t)/\tau$ can be obtained.

With the above proof, we can rewrite constraint (3.2) as follows,

$$\sum_{i=1}^{n} \varphi_i D_i^l(t) \leq f^{max} \tau. \tag{3.28}$$

After doing the above equivalent transformation, optimization problem (3.26) can be rewritten as,

$$\begin{aligned} \min_{\mathbf{D}^l(t), \mathbf{D}^r(t)} \quad & \{\frac{V\xi}{\tau^2} \sum_{i=1}^{n} [\varphi_i D_i^l(t)]^3 - \sum_{i=1}^{n} Q_i(t) D_i^l(t) \\ & + \sum_{i=1}^{n} [VP/R(t) - Q_i(t)] D_i^r(t) + VE^a(t)\}. \\ s.t. \quad & \sum_{i=1}^{n} D_i^r(t) \leq R(t)\tau. \\ & \sum_{i=1}^{n} \varphi_i D_i^l(t) \leq f^{max} \tau. \end{aligned} \tag{3.29}$$

The decision variables for Problem (3.29) are $\mathbf{D}^l(t)$ and $\mathbf{D}^r(t)$. At the same time, the two decision variables are decoupled. To facilitate the solution, we decompose it into two decoupled subproblems. The two subproblems are: local computation allocation and MEC computation allocation. Next, we will analyze each subproblem separately, and then give the optimal solution.

3.2.3 *Local Computation Allocation*

In this part, we will present the optimal solution to the local computation subproblem. The decision variable for the local computation subproblem is the amount of local computation. Therefore, we need to find the optimal local computation for mobile devices. Optimization problem (3.29) is decomposed into two subproblems. Here, we take the part containing the decision variable $D^l(t)$ as the local computation subproblem. Then, the optimal local computation allocation decision is obtained by solving this subproblem. The local computation subproblem is formulated as follows,

$$\min_{\mathbf{D}^l(t)} \frac{V\xi}{\tau^2}\sum_{i=1}^{n}[\varphi_i D_i^l(t)]^3 - \sum_{i=1}^{n} Q_i(t) D_i^l(t). \tag{3.30}$$

$$s.t. \sum_{i=1}^{n} \varphi_i D_i^l(t) \le f^{max}\tau.$$

It can be seen that Problem (3.30) is not easy to solve. Through analysis, we can know that if the value of $\sum_{i=1}^{n}\varphi_i D_i^l(t)$ has been determined, the solution to the problem can be easily solved. Therefore, we assume that the value of $\sum_{i=1}^{n}\varphi_i D_i^l(t)$ is deterministic. We can then simplify the problem to be solved. The simplified problem is formulated as follows

$$\min_{\mathbf{D}^l(t)} \; -\sum_{i=1}^{n}\frac{Q_i(t)}{\varphi_i}\varphi_i D_i^l(t). \tag{3.31}$$

The simplified problem is similar to the general minimum weight problem. Therefore, we solve it as a minimum weight problem. First, we can get the weight of the local computation $\varphi_i D_i^l(t)$ is $-\frac{Q_i(t)}{\varphi_i}$. Then, we give the optimal solution as follows,

$$D_i^l(t) = \begin{cases} \frac{\sum_{i=1}^{n}\varphi_i D_i^l(t)}{\varphi_i}, & i = i^* \\ 0, & otherwise, \end{cases} \tag{3.32}$$

where $i^* \in \text{argmax}_{i\in\{1,2,\cdots,n\}}\frac{Q_i(t)}{\varphi_i}$ denotes the set of applications with maximum $\frac{Q_i(t)}{\varphi_i}$. Therefore, when we determine the value of $\sum_{i=1}^{n}\varphi_i D_i^l(t)$, we can get the optimal local calculation amount according to (3.32). Actually, the value of $\sum_{i=1}^{n}\varphi_i D_i^l(t)$ is not really determined. Therefore, we solve the problem by the method of variable substitution. We replace $\sum_{i=1}^{n}\varphi_i D_i^l(t)$ with X. Then, we get the subproblem after rewriting as follows,

$$\min_{X} \frac{V\xi}{\tau^2}X^3 - Q_{i^*}(t)X. \tag{3.33}$$

$$s.t.\, 0 \le X \le f^{max}\tau.$$

The rewritten subproblem is very simple in form. We use the method of convex optimization to solve Problem (3.33). Then, the optimal solution X^* can be obtained by derivation of it. After we get the value of $X = \sum_{i=1}^{n}\varphi_i D_i^l(t)$, we can obtain the optimal local calculation amount according to the optimal solution (3.32).

3.2.4 MEC Computation Allocation

In this part, we will present the optimal solution to the MEC computation subproblem. The decision variable of the MEC computation subproblem is the offload computation. Therefore, we need to find the optimal offload computation for the mobile device. We decompose the MEC computation assignment subproblem from Problem (3.29). The MEC computation subproblem is formulated as follows,

$$\min_{\mathbf{D}^r(t)} \sum_{i=1}^{n}[VP/R(t) - Q_i(t)]D_i^r(t) + VE^a(t). \tag{3.34}$$

$$s.t. \sum_{i=1}^{n} D_i^r(t) \le R(t)\tau.$$

Equations (3.10) and (3.11) give the calculation methods for tail time and tail energy, respectively. Combining these two equations, we can see that there are two possible solutions for tail energy consumption. The equation below gives the calculation of the tail energy consumption.

$$E^a(t) = \begin{cases} e1, & \sum_{i=1}^{n} D_i^r(t) > 0 \\ e2, & otherwise. \end{cases} \tag{3.35}$$

It can be seen from Eq. (3.35) that the solution to the MCA subproblem is equal to $V \cdot e2$ when $\sum_{i=1}^{n} D_i^r(t) = 0$. We denote $V \cdot e2$ by $O1$. Then, when $\sum_{i=1}^{n} D_i^r(t) > 0$, we transform the MCA subproblem to (3.36).

$$\min_{\mathbf{D}^r(t)} \sum_{i=1}^{n}[VP/R(t) - Q_i(t)]D_i^r(t). \tag{3.36}$$

$$s.t. \ 0 < \sum_{i=1}^{n} D_i^r(t) \le R(t)\tau.$$

Similar to Problem (3.31), Problem (3.36) can also be viewed as a general minimum weight problem. First, we can conclude that the weight of offloading computation $D_i^r(t)$ is $VP/R(t) - Q_i(t)$. Then, we give the optimal solution as follows,

$$D_i^r(t) = \begin{cases} R(t)\tau, & i = i^* \\ 0, & otherwise, \end{cases} \tag{3.37}$$

Algorithm 2 Computation offloading and frequency scaling for energy efficiency (COFSEE)

1: Observe the current queue backlog of each application $Q_i(t)$.
2: Obtain the $\sum_{i=1}^{n} \varphi_i D_i^l(t)$ by solving the convex optimization problem (3.33).
3: **for all** $i \in I$ **do**
4: Calculate the $\frac{Q_i(t)}{\varphi_i}$ for the i-th application.
5: **end for**
6: **for all** $i \in I$ **do**
7: Search for index i^* with the maximum value of $\frac{Q_i(t)}{\varphi_i}$.
8: **end for**
9: Set the $D_i^l(t)$ according to (3.32).
10: Set the $f(t)$ according to (3.27).
11: Calculate the tail energy $e1$ and $e2$ according to (3.10) and (3.11).
12: **for all** $i \in I$ **do**
13: Calculate the $VP/R(t) - Q_i(t)$ for the i-th application.
14: **end for**
15: **for all** $i \in I$ **do**
16: Search for index i^* with the minimum value of $VP/R(t) - Q_i(t)$.
17: **end for**
18: Calculate the $O1$ and $O2$.
19: Set the $D_i^r(t)$ according to (3.39).

where $i^* \in \text{argmin}_{i \in \mathbf{N}}\{VP/R(t) - Q_i(t)\}$ denotes the set of applications with the smallest $VP/R(t) - Q_i(t)$. According to Eq. (3.37), we give the optimal solution $O2$. When $\sum_{i=1}^{n} D_i^r(t) > 0$, the optimal solution of MCA is shown below.

$$O2 = [VP/R(t) - Q_{i^*}(t)] \cdot R(t)\tau + V \cdot e1. \tag{3.38}$$

So far, we have analyzed two cases of MEC calculation allocation. Combining the above two situations, we give the optimal solution to the MCA problem as follows.

$$D_i^r(t) = \begin{cases} R(t)\tau, & i = i^*, O1 > O2 \\ 0, & i \neq i^*, O1 > O2 \\ 0, & O1 \leq O2. \end{cases} \tag{3.39}$$

In each time slot, we calculate the optimal local computation $\mathbf{D}^l(t)$ and the optimal offload computation $\mathbf{D}^r(t)$. Then, the task queue length of each application is updated according to Eq. (2.6). The specific content of the COFSEE algorithm is shown in Algorithm 2.

3.2.5 Theoretical Analysis

In this part, we focus on the performance of COFSEE. First we give the representation of the average energy consumption and average queue backlog of COFSEE.

We then prove the correctness of the given upper bound by derivation. Through theoretical analysis of COFSEE algorithm, we prove that COFSEE algorithm can guarantee the balance between energy consumption and queue backlog.

The average queue backlog for mobile devices is denoted by $\bar{Q}$ as follows,

$$\bar{Q} = \lim_{T\to\infty} \frac{1}{T} \sum_{t=0}^{T-1} \sum_{i\in\mathbf{N}} \mathbf{E}\{Q_i(t)\}. \tag{3.40}$$

Theorem 3.2 *Regardless of the value of V, when the task arrival rate of all applications of a given mobile device is $\boldsymbol{\lambda} = \{\lambda_1, \cdots, \lambda_n\}$. Assuming that if there is $\epsilon > 0$, and such that $\sum_{i=1}^{n}(\lambda_i + \epsilon) \in \Lambda$ holds, then we can get the upper limit of the average energy consumption of COFSEE, which is expressed as follows,*

$$\bar{E}^{our} \le e^* + \frac{C}{V}. \tag{3.41}$$

Not only that, we can also derive the average time queue backlog of COFSEE, which is expressed as follows,

$$\bar{Q} \le \frac{C + V(\hat{E} - \check{E})}{\epsilon}, \tag{3.42}$$

where C is a constant, given in Theorem 3.2, and e^* denotes the minimum average energy consumption.

The proof of Theorem 3.2 is given as follows.

In Theorem 3.2, we give an upper bound on the average energy consumption and the average queue backlog. Next, we will prove the correctness of the given upper bound. Therefore, we give Lemma 2.

Lemma 3.2 *For an arbitrary task arrival rate Λ. If $\sum_{i=1}^{n} \lambda_i \in \Lambda$, and there is an optimal task allocation decision $\boldsymbol{\pi}^* = \{\mathbf{D}^{l*}(t), \mathbf{D}^{r*}(t), f^*(t)\}$, and this decision is not affected by the queue length. And can perform task non-coordinated CPU cycle frequency scaling according to a fixed distribution. Then this optimal decision $\boldsymbol{\pi}^*$ can be expressed as follows.*

$$\mathbf{E}\{\bar{E}^{\boldsymbol{\pi}^*(t)}\} = e^*(\boldsymbol{\lambda}),$$

$$\mathbf{E}\{A_i(t)\} \le \mathbf{E}\{D_i^{l*}(t) + D_i^{r*}(t)\},$$

where $e^*(\boldsymbol{\lambda})$ represents the minimum average energy consumption under $\boldsymbol{\lambda} = \{\lambda_1, \cdots, \lambda_n\}$.

Proof Lemma 3.2 can be proved using Caratheodory's theorem. Here, for convenience, we do not give a complete proof.

In the above, we have defined the upper limit of the application's task arrival rate as A_i^{max}. Therefore, there will also be a range in the amount of energy consumed by processing and offloading tasks generated by these applications. We define the upper limit of energy consumption as $\hat{E}$ and the lower limit of energy consumption as $\check{E}$. Next, we use Lemma 3.2 to prove Theorem 3.41.

According to Lemma 3.2, we can know that if $\sum_{i=1}^{n}(\lambda_i + \epsilon) \in \Lambda$, and $\epsilon > 0$, an optimal strategy $\pi^\star$ can be obtained, and it satisfies the following,

$$\mathbf{E}\{\bar{E}^{\pi^\star(t)}\} = e^\star(\boldsymbol{\lambda} + \epsilon). \tag{3.43}$$

$$\mathbf{E}\{A_i(t)\} \le \mathbf{E}\{D_i^{l\star}(t) + D_i^{r\star}(t)\} - \epsilon. \tag{3.44}$$

Looking back at the previous article, we know that the goal of COFSEE is to minimize the upper bound of $\Delta_V(\Theta(t))$. Therefore, when given the optimal task assignment decision $\pi^\star$, we can get,

$$\begin{aligned}\Delta(\Theta(t)) + V\mathbf{E}\{E^{our}(t)|\Theta(t)\} &\le C + V\mathbf{E}\{e^{\pi\star}(t)|\Theta(t)\} \\ &+ \sum_{i\in\mathbf{N}} Q_i(t)\mathbf{E}\{A_i(t) - D_i^{l\star}(t) - D_i^{r\star}(t)|\Theta(t)\}.\end{aligned} \tag{3.45}$$

Combining (3.43) and (3.44), we can get the following inequality.

$$\begin{aligned}&\Delta(\Theta(t)) + V\mathbf{E}\{E^{our}(t)|\Theta(t)\} \\ &\qquad \le C + Ve^\star(\boldsymbol{\lambda} + \epsilon) - \epsilon \sum_{i\in\mathbf{N}} Q_i(t).\end{aligned} \tag{3.46}$$

Adding both sides of Eq. (3.46) iteratively simultaneously from time slot 0 with time slot $t-1$, we can get the following,

$$\begin{aligned}&\mathbf{E}\{L(\Theta(T))\} - \mathbf{E}\{L(\Theta(0))\} + V\sum_{t=0}^{T-1}\mathbf{E}\{E^{our}(t)\} \\ &\le CT + VTe^\star(\boldsymbol{\lambda} + \epsilon) - \epsilon\sum_{t=0}^{T-1}\sum_{i\in\mathbf{N}}\mathbf{E}\{Q_i(t)\}.\end{aligned} \tag{3.47}$$

In general, when $t = 0$, the application has not yet generated tasks, the queue length is 0. Therefore, we can get $L(\Theta(0)) = 0$. Then again, because $Q_i(T)$ and $\mathbf{E}\{L(\Theta(T))\}$ are both non-negative. Therefore, we have,

$$V\sum_{t=0}^{T-1}\mathbf{E}\{E^{our}(t)\} \le CT + VTe^{\star}(\boldsymbol{\lambda}+\epsilon) - \epsilon\sum_{t=0}^{T-1}\sum_{i\in\mathbf{N}}\mathbf{E}\{Q_i(t)\}. \tag{3.48}$$

Because ϵ is non-negative and satisfies $\epsilon\sum_{t=0}^{T-1}\sum_{i\in\mathbf{N}}\mathbf{E}\{Q_i(t)\}\ge 0$. Therefore, dividing both sides of Eq. (3.48) by VT, we get,

$$\frac{1}{T}\sum_{t=0}^{T-1}\mathbf{E}\{E^{our}(t)\} \le \frac{C}{V} + e^{\star}(\boldsymbol{\lambda}+\epsilon). \tag{3.49}$$

When $\epsilon \to 0$, $T \to \infty$, apply the Lebesgues dominated convergence theorem, we obtain (3.41).

According to (3.46), we also obtain

$$\epsilon\sum_{t=0}^{T-1}\sum_{i\in\mathbf{N}}\mathbf{E}\{Q_i(t)\} \le CT + VTe^{\star}(\boldsymbol{\lambda}+\epsilon) - V\sum_{t=0}^{T-1}\mathbf{E}\{E^{our}(t)\}. \tag{3.50}$$

Earlier, we have given upper and lower bounds on energy consumption. Therefore, we can get,

$$\frac{1}{T}\sum_{t=0}^{T-1}\sum_{i\in\mathbf{N}}\mathbf{E}\{Q_i(t)\} \le \frac{C+V(\hat{E}-\check{E})}{\epsilon}. \tag{3.51}$$

When $T \to \infty$, we obtain (3.42).

Remark In Theorem 3.2, we give an upper bound on the average energy consumption and average queue length of COFSEE. From Theorem 3.2, we can know that the average energy consumption decreases as the value of V increases, while the average queue length increases as the value of V increases. As the average queue length increases with the value of V, it still has an upper limit. Through analysis, COFSEE can achieve a balance between energy consumption and queue length. It can be seen from Eq. (3.41) that when the value of V is large enough, the upper limit of the average energy consumption is equal to e^*, that is, COFSEE can theoretically achieve the optimal energy consumption. But the larger the value of V, the larger the average queue length. Therefore, in order not to make the queue backlog too large, we can set a suitable value of V to fit our system.

The algorithmic details of COFSEE are given in Algorithm 2. Next, we will give the time complexity of the COFSEE algorithm. In Algorithm 2, there are two inner loops (lines 6–8 and 15–17). The specific flow of the algorithm is as follows. First, solve the convex optimization problem (3.33) to obtain $\sum_{i=1}^{n} \varphi_i D_i^l(t)$. Second, according to Eq. (3.32), the optimal local computation amount for each application is obtained. Then the tail energy consumption $e1$ and $e2$ are obtained. Finally, according to Eq. (3.39), the optimal offload computation amount for each application is obtained. Since there are only n applications in total, and COFSEE traverses each application once in each loop. Therefore, we can get the time complexity of COFSEE to be $O(n)$. According to the above analysis and proof, a conclusion can be drawn. The COFSEE algorithm is also feasible for large systems.

3.3 Performance Evaluation

In this part, we will evaluate the parameters of the COFSEE algorithm to judge the reliability of our algorithm. Then, we give comparative experiments to judge the effectiveness of our algorithm. Next, we give specific information on the experimental setup.

In this experiment, we consider a system consisting of a mobile device and an edge server. Then, there are n applications running on this mobile device. In our experiments, there are three types of applications running on the device. Types of apps include video transcoding, board games, and 6-queues puzzle games. The tasks generated by the application are partially processed locally and partially transferred to the server for processing. The task amount $A_i(t)$ generated by each application in each time slot satisfies a uniform distribution, that is, $A_i(t) \sim U[0, 2200]$ bits. We know that the COFSEE algorithm does not need to know the information of task generation and arrival. The number of CPU cycles required by the application to process 1 bit data satisfies a uniform distribution, that is, $\varphi_i \sim U[1000, 2000]$ cycles/bit. The wireless channel is considered as a small-scale Ralyigh fading model, and the channel power gain $h_i(t)$ follows an exponential distribution with unit mean, i.e., $h_i(t) \sim E(1)$.

In addition, the time slot length τ is set to 1 s, the transmit power P of the mobile device is set to 1.6 W, the channel bandwidth B is set to 1 MHz, the channel noise power σ^2 is set to 10^{-6} W, and the maximum CPU cycle frequency f^{max} of the mobile device is set to 1 GHz. The tail signal power P_T is set to 1.1 W, when there is no data transmission, the time δ_T that the channel stays in the TA phase is set to 10 s, and the effective switched capacitor ξ is set to 10^{-27}. The parameter settings are summarized in Table 3.2. Each experiment was run 3000 times and then averaged.

Table 3.2 Simulation parameters

Parameters	Values
The task amount generated by each application i per time slot	$U[0, 2200]$ (bit)
The number of CPU cycles required by the application to process 1 bit data	$U[1000, 2000]$ (cycles/bit)
The channel power gain	$E(1)$
The time slot length	1 s
The transmit power of the mobile device	1.6 W
The channel noise power	10^{-6} W
The maximum CPU cycle frequency of the mobile device	1 GHz
The tail signal power	1.1 W
The time that the channel stays in the TA phase	10 s
The effective switched capacitor	10^{-27}

3.3.1 Impacts of System Parameters

3.3.1.1 Effect of Tradeoff Parameter V

Figures 3.1 and 3.2 show how the energy consumption and queue length vary with the parameter V, respectively. In the COFSEE algorithm, the parameter V is used to balance energy consumption and queue length.

Figure 3.1 shows the change in energy consumption at different V values. It can be seen from the figure that there is a negative correlation between the energy consumption and the parameter V. With the increase of V, the energy consumption decreases gradually. Because as the value of V increases, the weight of energy consumption also increases. Therefore, in order to ensure the stability of mobile devices, the COFSEE algorithm will dynamically adjust the task allocation decisions of applications to reduce system energy consumption. This result shown in the figure is consistent with (3.41) in Theorem 3.2.

Figure 3.2 shows the variation of queue length at different V values. It can be seen from the figure that as the value of V increases, the queue length also increases gradually. However, the queue length will not increase all the time, but will gradually converge to a balanced state. This result shown in the figure is consistent with (3.42) in Theorem 3.2. Combining Figs. 3.1 and 3.2, we can see that the COFSEE algorithm can guarantee the stability of the device regardless of the value of V. Moreover, we can obtain the optimal task allocation decision by changing the value of V and ensure the stability of the queue length. At the same time, when the value of V is larger, the energy consumed will be smaller.

3.3.1.2 Effect of Arrival Rate

Figures 3.3 and 3.4 show the variation of energy consumption and queue length with different task arrival rates, respectively. In our experiments, we set the task

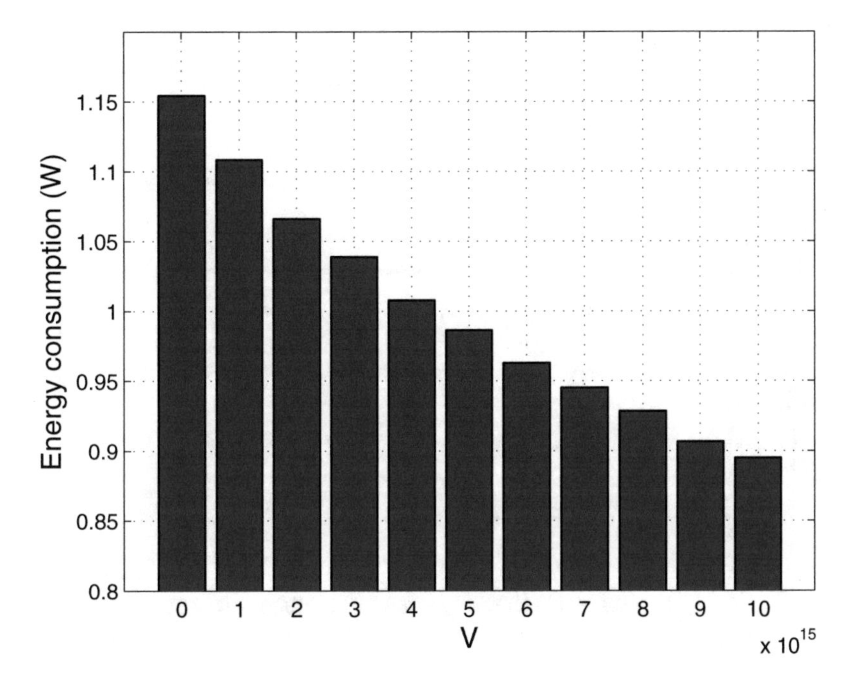

Fig. 3.1 Energy consumption with different values of V

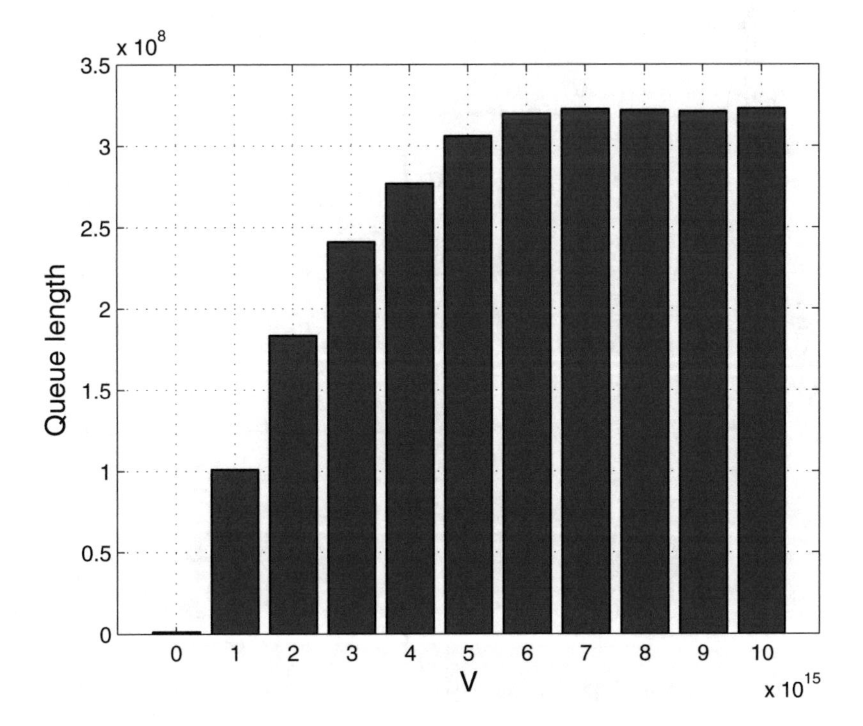

Fig. 3.2 Queue length with different values of V

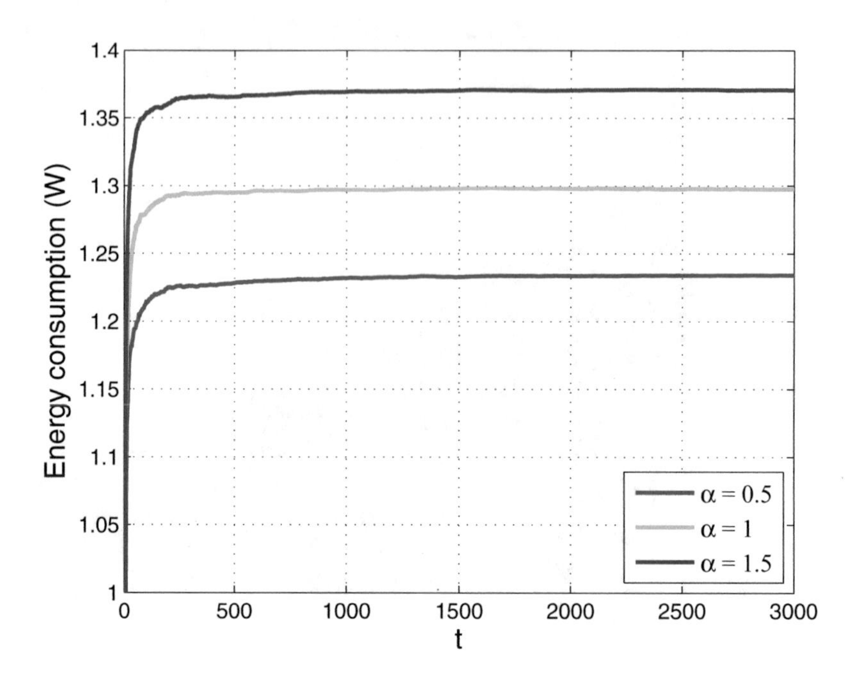

Fig. 3.3 Energy consumption with different arrival rate

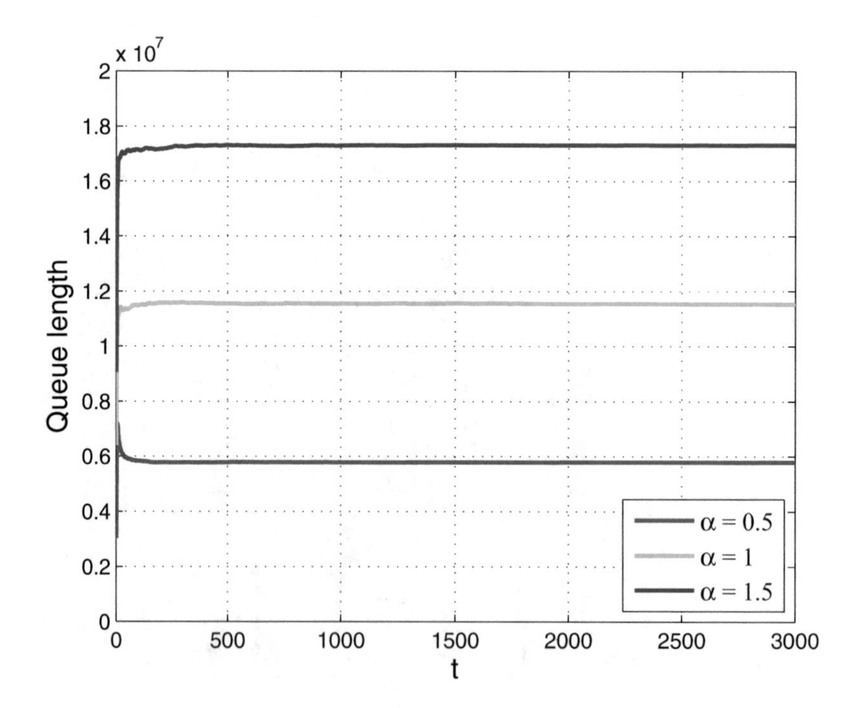

Fig. 3.4 Queue length with different arrival rate

arrival rate for each application to $\alpha \cdot A_i(t)$, where the value of α is 0.5, 1, and 1.5, respectively.

Figure 3.3 shows the effect of different task arrival rates on energy consumption. It can be seen from the figure that for different task arrival rates, the system energy consumption increases rapidly at first, and then gradually stabilizes. However, in the same time slot, the higher the task arrival rate, the higher the energy consumption. This is because the larger the task arrival rate, the more tasks the application generates per time slot. However, although the task arrival rate is different, the data transfer rate and local computing power do not change. When the task arrival rate is larger, more computing tasks will be allocated to local computing and offloaded. Therefore, more energy consumption will be generated. When the amount of computation generated reaches the local processing capacity and the offload transmission capacity, COFSEE will dynamically adjust the task allocation decision to ensure the stability of energy consumption.

Figure 3.4 shows the effect of different task arrival rates on queue length. It can be seen from the figure that for different task arrival rates, the queue length increases rapidly at first, and then gradually stabilizes. However, within the same time slot, the higher the task arrival rate, the larger the queue length. This is because the higher the task arrival rate, the more tasks are generated in each slot. However, despite the different task arrival rates, neither the local processing power nor the offload transfer power changes. Therefore, when the task arrival rate is higher, the backlog of tasks is larger, and the queue length is larger. It can be seen from the figure that the queue lengths will converge under different task arrival rates. This result shows that the COFSEE algorithm can dynamically adjust the task assignment decision according to the different task arrival rates.

3.3.1.3 Effect of Slot Length

Figure 3.5 shows the variation of queue length under different slot lengths. In our experiments, we set the time slot length as τ, where the value of τ is 0.5 s, 1 s, and 1.5 s, respectively. It can be seen from the figure that, for different time slot lengths, the queue length increases rapidly at first, and then tends to stabilize gradually. However, within the same time slot, if the time slot length is larger, the queue length will also be larger. This is because the larger the time slot length, the more tasks each application generates. However, the local processing power of the mobile device and the data transfer rate of the channel do not change. Therefore, more tasks are backlogged, making the queue length larger. At the same time, we can see that the larger the slot length, the more difficult it is for the COFSEE algorithm to adapt to the system dynamics. The smaller the time slot length is, the more expensive the COFSEE algorithm is to obtain system parameters or variables.

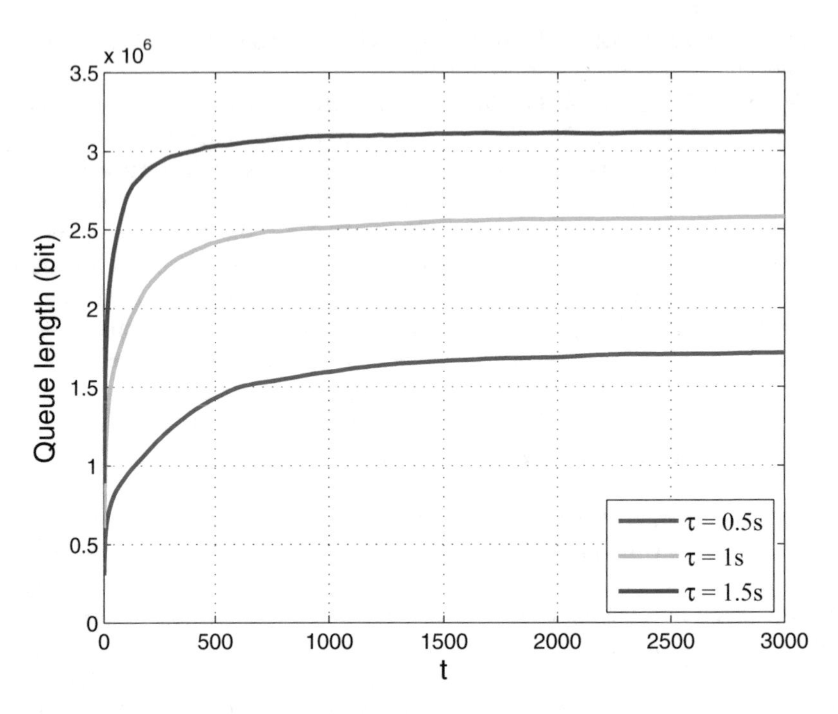

Fig. 3.5 Queue length with different slot lengths

3.3.2 Performance Comparison with RLE, RME and TS Schemes

In order to better evaluate the performance of the COFSEE algorithm and prove that COFSEE algorithm can better solve the problem of computation offloading, we compared the COFSEE algorithm with Round-robin Local Execution algorithm, Round-robin MEC Execution algorithm and task scheduling algorithm.

- **Round-robin Local Execution (RLE)** : In this algorithm, the computing tasks generated by each application are all executed locally. When dealing with these tasks, all applications need to be sorted. Only one application-generated task is processed per time slot. The tasks of different applications are processed sequentially in sequence.
- **Round-robin MEC Execution (RME)** : In this algorithm, the computing tasks generated by each application are offloaded for execution. When dealing with these tasks, all applications need to be sorted. Each time slot only processes the tasks generated by one application, and the tasks of different applications are processed in sequence.

Figure 3.6 shows the relationship of energy consumption between three different algorithms. Figure 3.7 shows the relationship of queue lengths between three dif-

Table 3.3 Execution time of different algorithms

Algorithm	Execution time (ms)
COFSEE	0.7760
RLE	0.0166
RME	0.0204

ferent algorithms. As can be seen from the figure, our algorithm can simultaneously minimize the transmission energy consumption and the queue length.

As can be seen from Fig. 3.6, the energy consumption of the COFSEE algorithm is the smallest among the three algorithms. Specifically, the average energy consumption of the COFSEE algorithm is about 15% lower than that of the RLE algorithm. At the same time, the average energy consumption of the COFSEE algorithm is about 38% lower than that of the RME algorithm. Experiments show that our algorithm has a greater advantage in reducing device energy consumption. This is because our algorithm can adapt to the dynamic change of the channel, that is, the COFSEE algorithm can detect the change of the channel in time and respond quickly. Then dynamically adjust the application's task allocation strategy and CPU cycle frequency to ensure that energy consumption is minimized.

As can be seen from Fig. 3.7, the queue length of the RLE algorithm increases linearly. Therefore, the RLE algorithm can cause the mobile device to become unstable. This is because the local processing capacity of the mobile device is limited, however, the amount of newly arrived tasks per time slot exceeds the upper limit that can be processed locally. As a result, computing tasks cannot be processed in a timely manner, resulting in a continuous backlog of computing tasks on mobile devices, resulting in an increasing queue length. At the same time, it can also be seen from the figure that the queue lengths of the COFSEE algorithm and the RME algorithm are almost the same and very small. This is because the idea of the RME algorithm is to offload tasks to the MEC server for processing. Therefore, the queue length of the RME algorithm becomes very small. Experiments show that our algorithm can achieve low queue backlog and ensure the stability of mobile devices. Combining with Figs. 3.6 and 3.7, it can be seen that the COFSEE algorithm can effectively reduce energy consumption and queue length, and maintain the stability of mobile devices.

The execution time of the three algorithms is shown in Table 3.3. In our experiments, each algorithm was run 500 times and the average result was calculated. As can be seen from Table 3.3, the execution time of the RLE algorithm is the smallest, followed by the RME algorithm, and the execution time of the COFSEE algorithm is the largest. This is because the COFSEE algorithm needs to pay attention to system information such as wireless channel status, queue length and tail time when running. Then, collect and organize this information to make optimal task allocation decisions. Although the execution time of the COFSEE algorithm is the largest, it is only 0.776 s, which is within the acceptable range. Therefore, in general, the COFSEE algorithm is relatively optimal.

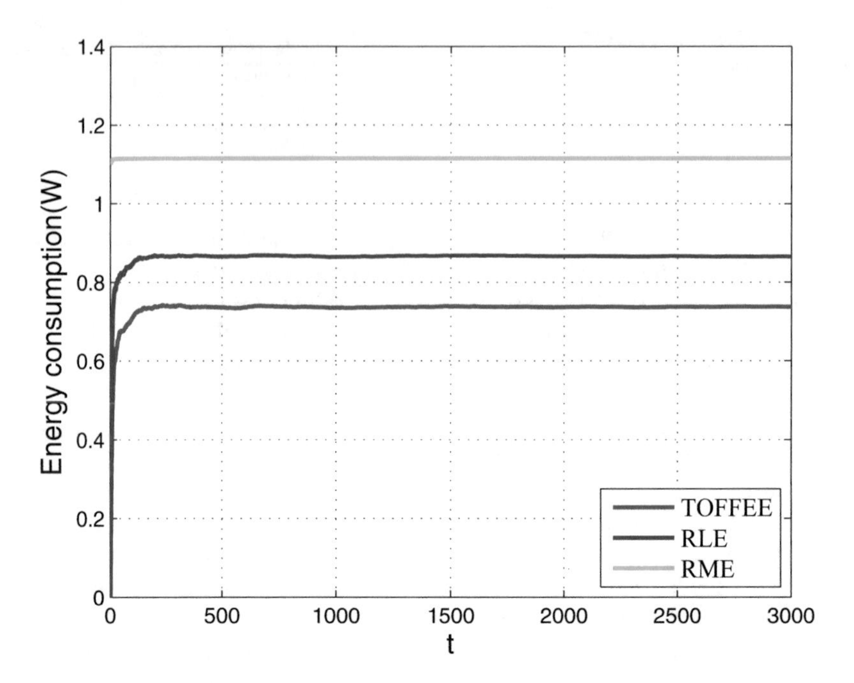

Fig. 3.6 Energy consumption with the three different algorithms

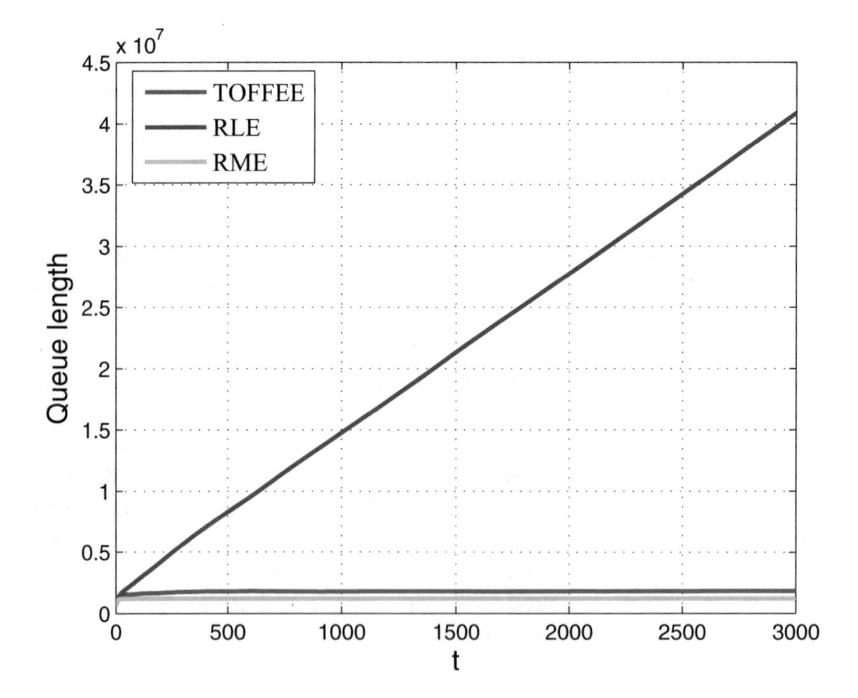

Fig. 3.7 Queue length with the three different algorithms

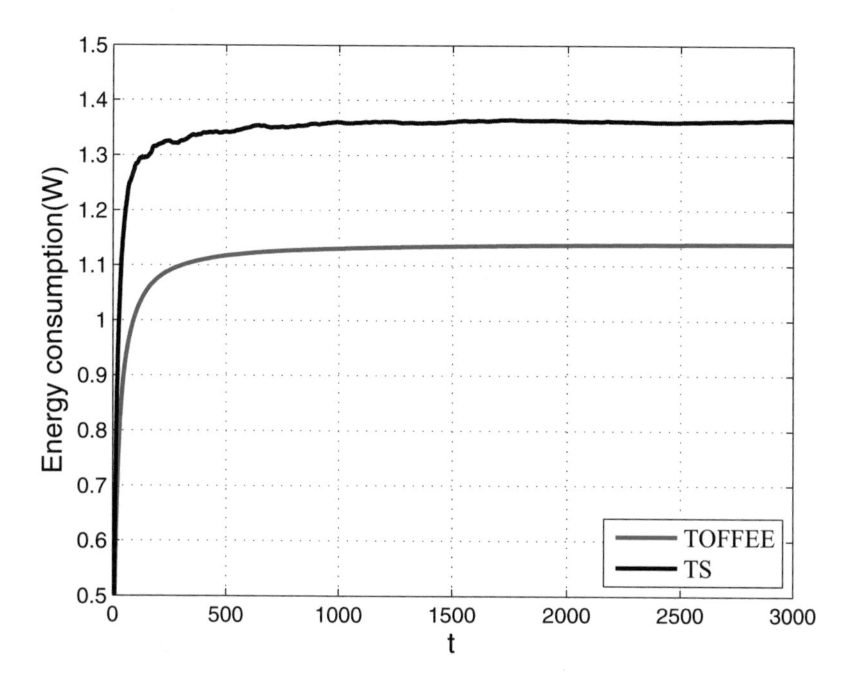

Fig. 3.8 Energy consumption of COFSEE and TS

Although the above experiments prove that our algorithm has good performance, it is not comprehensive. Therefore, we compare the COFSEE algorithm with the task scheduling algorithm to further evaluate the performance of COFSEE. Figure 3.8 shows the energy consumption relationship between the COFSEE algorithm and the TS algorithm. As can be seen from the figure, the energy consumption of the TS algorithm is greater than that of the COFSEE algorithm. This is because we comprehensively consider the local computing energy consumption, data transmission energy consumption and tail energy consumption, and optimize them simultaneously. Figure 3.9 shows the queue length relationship between the COFSEE algorithm and the TS algorithm. As can be seen from the figure, the queue length of the TS algorithm is larger than that of the COFSEE algorithm. This is because the COFSEE algorithm can jump the task allocation decision in time according to the change of the channel state to avoid the backlog of the task queue.

3.4 Literature Review

In recent years, with the increase of computation-intensive applications, more and more computing resources are required for mobile devices [33, 34]. However, due to the limitations of the mobile device itself, such as limited battery capacity,

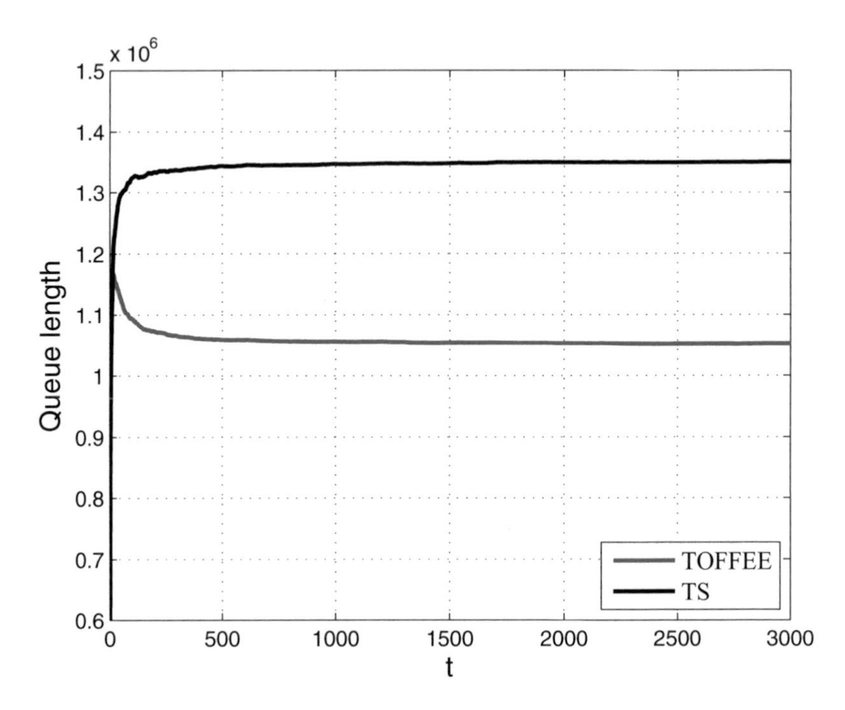

Fig. 3.9 Queue length of COFSEE and TS

limited processing speed, etc., the task processing cannot be performed well. Therefore, the MEC model came into being. MEC brings great convenience to mobile devices. With the development of IoT technology, MEC has received more extensive attention. Sun et al. [35] studied the mobility management problem in the combination of MEC and base station. They proposed a user-centric energy-aware mobility scheme. The scheme combined Lyapunov optimization theory and avoidance theory, and did not require statistical system information. The scheme realized delay optimization and wireless access in an online manner and under energy constraints. They experimentally demonstrated that this scheme could achieve optimal latency performance under energy constraints.

Dinh et al. [11] proposed a task offloading optimization framework that enabled offloading from a single mobile device to multiple edge devices. To make this optimization framework more extensive, both cases when the CPU frequency was fixed and when the CPU frequency was varied were investigated. They considered both task allocation decisions and CPU cycle frequency, and then analyzed the task's total execution latency and mobile device energy consumption. And it was proved by experiments that the framework could achieve improved performance.

Neto et al. [36] proposed a user-based online offloading framework. The framework contained a decision component. Therefore, the system overhead could be reduced to a certain extent during the system operation, and a better user experience could be brought to the user. He et al. [37] studied the problem of request scheduling in MEC. They studied both the allocation of shareable resources and the allocation of non-shareable resources. At the same time, an effective heuristic algorithm was designed. The algorithm could achieve near-optimal performance under the constraints of communication, computation and storage.

Dai et al. [38] studied the MEC-based computation offloading problem. They designed a novel two-layer computation offloading framework. In order to minimize system energy consumption, they jointly solved the problem of computation offloading and user association. Then, they devised an algorithm. Experiments showed that the algorithm was efficient in computing resource allocation and transmission power allocation. Chen et al. [39] studied the problem of efficient computation offloading in mobile edge cloud computing. In order to allow users to perform task offloading more flexibly, they combined the user's task offloading with the resource allocation of the network operator. To solve this NP-hard problem, they devised an efficient approximate solution. And the effectiveness of on-demand task offloading was proved by experiments.

In order to improve the capability of mobile edge computing, Ma et al. [40] proposed a cloud-assisted mobile computing framework, and modeled the system as a queuing network. To balance system delay and cost, a minimization problem was proposed. They designed a linear algorithm to solve this convex optimization problem. Finally, the simulation experiment proved that the computing framework could achieve a balance between energy consumption and delay. Xiao et al. [41] studied the workload problem in fog computing networks, using fog nodes to share part of the work offload to improve user experience quality. They considered the quality of user experience and the power of fog nodes, and proposed a cloud computing network with multi-fog node coordination. Then, they proposed a distributed optimization algorithm based on alternating method multipliers to maximize user experience quality. They experimentally demonstrated the performance of the proposed framework, which could improve system performance.

Although the above work has achieved the improvement of system performance and the stability of the system. However, these works are based on the premise that the wireless channel state or the arrival information of computing tasks can be known in advance. However, we know that in practical applications, the arrival of tasks is uncontrollable [42, 43]. We cannot enforce when tasks are generated, how they are generated, and how many tasks are generated. Due to the influence of weather or human factors, the channel state is also constantly changing [44]. Therefore, it is difficult for us to accurately predict the arrival of tasks and the state of wireless channels in advance. Therefore, the dynamics of the actual environment should be considered when conducting research on computation offloading strategies. The above work mainly focuses on performance over time. But in our actual life, we can't just use it for a certain period of time. What we need to focus on is long-term system change. Therefore, in order to better meet the actual needs, a lot of work has begun to focus on the long-term performance of the system. We also propose a stochastic optimization problem for highly dynamically changing task arrivals and wireless networks.

The MEC computing model is a model emerging to meet the growing computing demands. Mobile devices can offload computation-intensive tasks to the MEC to relieve the pressure on mobile devices. However, the battery capacity of mobile devices is limited, which may affect the quality of user experience. Therefore, new edge computing methods based on energy harvesting technology have emerged [45–47]. Mao et al. [20] investigated green edge computing systems with energy harvesting devices. They considered a green MEC system under a single user. Then, an online computation offloading algorithm based on Lyapunov optimization was proposed. They also considered task offloading decisions, offload transfer power, and CPU cycle frequency. They conducted simulation experiments to verify the effectiveness of the proposed algorithm, which could effectively minimize the execution cost.

Min et al. studied the problem of task offloading with energy harvesting devices. They proposed an energy harvesting offloading scheme based on reinforcement learning. The scheme did not need to know about the MEC model and the computation delay model. We also study issues such as energy consumption, computation latency, and utility. Simulation experiments show that this scheme can effectively improve the utility of IoT devices. Guo et al. [48] studied the energy harvesting problem and caching problem in small cell networks. They designed an optimization problem that combined energy harvesting and caching. In order to facilitate the solution of the problem, they decomposed it. Then, the optimization problem was solved by game theory and iterative methods. The effectiveness of the proposed algorithm was verified by simulation experiments.

Wu et al. [49] investigated the high intermittency and unpredictability of energy harvesting in MEC. To address these new challenges, they designed an algorithm to optimize system performance. The algorithm could combine load balancing and admission control. Then, they solved the energy causal constraint problem using Lyapunov optimization techniques. Since the designed algorithm didn't require any prior information, a near-optimal system performance was achieved. Qian et al. [50] studied the hybrid energy problem in small heterogeneous cellular networks. They focused on the problem of system revenue maximization and power allocation. To solve this convex optimization problem, they proposed a power allocation algorithm based on dual parameters. They verified that the proposed algorithm could better cope with fast fading channels through simulation experiments.

Zhang et al. [51] studied the energy consumption problem in a multi-user MEC system. They considered mobile devices with energy harvesting devices. They formulated the problem of minimizing system energy consumption as a stochastic optimization problem. Then, this problem was solved using Lyapunov optimization technique. At the same time, a distributed online algorithm was designed, and the effectiveness of the proposed algorithm was proved by experiments. Yang et al. [20] studied the problem of computation offloading in 5G, considering the task offloading of cell networks. They jointly considered task computation and device communication, and modeled the energy consumption of task offloading as a delay-constrained optimization problem. The research goal of this problem was to minimize the total energy consumption of the system. Inspired by the fish swarm algorithm, they proposed an optimized offloading scheme based on this. The effectiveness of the proposed scheme was verified by simulation experiments.

You et al. [52] studied the resource allocation problem of mobile edge computation offloading. They modeled the resource allocation problem as a convex optimization problem with computation latency as a constraint. Then an offload priority function was designed, which could combine channel gain and local computing energy consumption to determine user offload priority, so as to minimize the energy consumption of mobile devices. Zhang et al. [53] proposed an energy-aware offloading scheme. They proposed a system model that integrated computation offloading and resource allocation, and jointly optimized computation offloading and resource allocation to achieve a balance between energy consumption and latency.

The above work only considers the energy consumption of data transmission when calculating the energy consumption of the system. None of these works consider tail energy consumption. However, tail energy consumption is also very important in computation offloading [54]. In particular, with the development of IoT technology, there are more and more computation-intensive applications, so how to achieve more efficient offloading is of great concern. Therefore, we cannot ignore tail energy consumption and should pay attention to it. Therefore, there are also some works expounding the importance of tail energy consumption [18, 55]. In order to be more in line with the actual needs, we also consider the tail energy consumption. At the same time, we consider the transmission energy consumption and tail energy consumption together. in order to optimize them simultaneously.

3.5 Summary

This chapter studies the task assignment problem of MEC system. To better determine the optimal task assignment decision, we jointly consider task assignment and local CPU cycle frequency. In the calculation of system energy consumption, we introduce tail energy consumption in order to make the calculation of energy consumption more reasonable. We transform the problem of minimizing energy consumption into a deterministic optimization problem using Lyapunov optimization techniques. At the same time, we need to ensure the upper limit of the task queue length to ensure the stability of mobile devices. Using Lyapunov optimization techniques can simplify the problem because we do not need system priors. Then, we design a stochastic optimization offloading algorithm with polynomial time complexity. Since the algorithm does not need the information of task arrival and channel state, the design of the algorithm becomes simple and easy to operate. Through analysis, COFSEE algorithm can achieve the trade-off between system energy consumption and task queue length by changing the value of parameter V. We can also increase the value of V to reduce system energy consumption. We experimentally demonstrate that the algorithm can achieve a balance between energy consumption and queue length. At the same time, we also conducted comparative experiments. The experimental results show that our algorithm can minimize the system energy consumption while ensuring a short queue length.

References

1. D. Chen et al., S2M: a lightweight acoustic fingerprints-based wireless device authentication protocol. IEEE Int. Things J. **4**(1), 88–100 (2017). https://doi.org/10.1109/JIOT.2016.2619679
2. P. Dai et al., Multi-armed bandit learning for computation-intensive services in MEC-empowered vehicular networks. IEEE Trans. Veh. Technol. **69**(7), 7821–7834 (2020). https://doi.org/10.1109/TVT.2020.2991641
3. Q. Kuang, J. Gong, X. Chen, X. Ma, Analysis on computation-intensive status update in mobile edge computing. IEEE Trans. Veh. Technol. **69**(4), 4353–4366 (2020). https://doi.org/10.1109/TVT.2020.2974816
4. F. Liu, P. Shu, J.C.S. Lui, AppATP: an energy conserving adaptive mobile-cloud transmission protocol. IEEE Trans. Comput. **64**(11), 3051–3063 (2015). https://doi.org/10.1109/TC.2015.2401032
5. H. Qi, A. Gani, Research on mobile cloud computing: review, trend and perspectives, in *2012 Second International Conference on Digital Information and Communication Technology and it's Applications (DICTAP)* (2012), pp. 195–202. https://doi.org/10.1109/DICTAP.2012.6215350
6. L.A. Tawalbeh, R. Mehmood, E. Benkhlifa, H. Song, Mobile cloud computing model and big data analysis for healthcare applications. IEEE Access **4**, 6171–6180 (2016). https://doi.org/10.1109/ACCESS.2016.2613278
7. X. Wang et al., Dynamic resource scheduling in mobile edge cloud with cloud radio access network. IEEE Trans. Parallel Distrib. Syst. **29**(11), 2429–2445 (2018). https://doi.org/10.1109/TPDS.2018.2832124
8. C. Wang, C. Liang, F.R. Yu, Q. Chen, L. Tang, Computation offloading and resource allocation in wireless cellular networks with mobile edge computing. IEEE Trans. Wirel. Commun. **16**(8), 4924–4938 (2017). https://doi.org/10.1109/TWC.2017.2703901
9. X. Hu, K.-K. Wong, K. Yang, Wireless powered cooperation-assisted mobile edge computing. IEEE Trans. Wirel. Commun. **17**(4), 2375–2388 (2018). https://doi.org/10.1109/TWC.2018.2794345
10. Y. Mao, C. You, J. Zhang, K. Huang, K.B. Letaief, A survey on mobile edge computing: the communication perspective. IEEE Commun. Surv. Tutorials **19**(4), 2322–2358 (2017). https://doi.org/10.1109/COMST.2017.2745201
11. T.Q. Dinh, J. Tang, Q.D. La, T.Q.S. Quek, Offloading in mobile edge computing: task allocation and computational frequency scaling. IEEE Trans. Commun. **65**(8), 3571–3584 (2017). https://doi.org/10.1109/TCOMM.2017.2699660
12. Y. Jararweh, A. Doulat, O. AlQudah, E. Ahmed, M. Al-Ayyoub, E. Benkhelifa, The future of mobile cloud computing: integrating cloudlets and mobile edge computing, in *2016 23rd International Conference on Telecommunications (ICT)* (2016), pp. 1–5. https://doi.org/10.1109/ICT.2016.7500486
13. H. Wu, Y. Sun, K. Wolter, Energy-efficient decision making for mobile cloud offloading. IEEE Trans. Cloud Comput. **8**(2), 570–584 (2020). https://doi.org/10.1109/TCC.2018.2789446
14. M. Jia, J. Cao, L. Yang, Heuristic offloading of concurrent tasks for computation-intensive applications in mobile cloud computing, in *2014 IEEE Conference on Computer Communications Workshops (INFOCOM WKSHPS)* (2014), pp. 352–357. https://doi.org/10.1109/INFCOMW.2014.6849257
15. M.A. Hassan, Q. Wei, S. Chen, Elicit: efficiently identify computation-intensive tasks in mobile applications for offloading, in *2015 IEEE International Conference on Networking, Architecture and Storage (NAS)* (2015), pp. 12–22. https://doi.org/10.1109/NAS.2015.7255215

16. Y. Mao, J. Zhang, K.B. Letaief, Dynamic computation offloading for mobile-edge computing with energy harvesting devices. IEEE J. Sel. Areas Commun. **34**(12), 3590–3605 (2016). https://doi.org/10.1109/JSAC.2016.2611964
17. Y. Mao, J. Zhang, S.H. Song, K.B. Letaief, Stochastic joint radio and computational resource management for multi-user mobile-edge computing systems. IEEE Trans. Wirel. Commun. **16**(9), 5994–6009 (2017). https://doi.org/10.1109/TWC.2017.2717986
18. Y. Cui et al., Performance-aware energy optimization on mobile devices in cellular network. IEEE Trans. Mob. Comput. **16**(4), 1073–1089 (2017). https://doi.org/10.1109/TMC.2016.2586052
19. M.E.T. Gerards, J.L. Hurink, J. Kuper, On the interplay between global DVFS and scheduling tasks with precedence constraints. IEEE Trans. Comput. **64**(6), 1742–1754 (2015). https://doi.org/10.1109/TC.2014.2345410
20. L. Yang, H. Zhang, M. Li, J. Guo, H. Ji, Mobile edge computing empowered energy efficient task offloading in 5G. IEEE Trans. Veh. Technol. **67**(7), 6398–6409 (2018). https://doi.org/10.1109/TVT.2018.2799620
21. Y. Hu, T. Cui, X. Huang, Q. Chen, Task offloading based on lyapunov optimization for MEC-assisted platooning, in *2019 11th International Conference on Wireless Communications and Signal Processing (WCSP)* (2019), pp. 1–5. https://doi.org/10.1109/WCSP.2019.8928035
22. J. Yang, Q. Yang, K.S. Kwak, R.R. Rao, Power–delay tradeoff in wireless powered communication networks. IEEE Trans. Veh. Technol. **66**(4), 3280–3292 (2017). https://doi.org/10.1109/TVT.2016.2587101
23. Y. Sun, T. Wei, H. Li, Y. Zhang, W. Wu, Energy-efficient multimedia task assignment and computing offloading for mobile edge computing networks. IEEE Access **8**, 36702–36713 (2020). https://doi.org/10.1109/ACCESS.2020.2973359
24. W. Hu, G. Cao, Quality-aware traffic offloading in wireless networks. IEEE Trans. Mob. Comput. **16**(11), 3182–3195 (2017). https://doi.org/10.1109/TMC.2017.2690296
25. Y. Chen, N. Zhang, Y. Zhang, X. Chen, W. Wu, X.S. Shen, TOFFEE: task offloading and frequency scaling for energy efficiency of mobile devices in mobile edge computing. IEEE Trans. Cloud Comput. **9**(4), 1634–1644 (2021). https://doi.org/10.1109/TCC.2019.2923692
26. K. Wang, K. Yang, C.S. Magurawalage, Joint energy minimization and resource allocation in C-RAN with mobile cloud. IEEE Trans. Cloud Comput. **6**(3), 760–770 (2018) . https://doi.org/10.1109/TCC.2016.2522439
27. F. Wang, J. Xu, X. Wang, S. Cui, Joint offloading and computing optimization in wireless powered mobile-edge computing systems. IEEE Trans. Wirel. Commun. **17**(3), 1784–1797 (2018). https://doi.org/10.1109/TWC.2017.2785305
28. A. Khlass, D. Laselva, R. Jarvela, On the flexible and performance-enhanced radio resource control for 5G NR networks, in *2019 IEEE 90th Vehicular Technology Conference (VTC2019-Fall)* (2019), pp. 1–6. https://doi.org/10.1109/VTCFall.2019.8891551
29. Y. Zhou, F. Tang, Y. Kawamoto, N. Kato, Reinforcement learning-based radio resource control in 5G vehicular network. IEEE Wirel. Commun. Lett. **9**(5), 611–614 (2020). https://doi.org/10.1109/LWC.2019.2962409
30. J. Huang, F. Qian, A. Gerber, Z.M. Mao, S. Sen, O. Spatscheck, A close examination of performance and power characteristics of 4g lte networks, in *International Conference on Mobile Systems, Applications, and Services* (2012), pp. 225–238
31. M.J. Neely, *Stochastic Network Optimization With Application to Communication and Queueing Systems* (Morgan & Claypool, Williston, 2010)
32. D. Zhang et al., Resource allocation for green cloud radio access networks with hybrid energy supplies. IEEE Trans. Veh. Technol. **67**(2), 1684–1697 (2018). https://doi.org/10.1109/TVT.2017.2754273
33. X. Yang, X. Yu, H. Huang, H. Zhu, Energy efficiency based joint computation offloading and resource allocation in multi-access MEC systems. IEEE Access **7**, 117054–117062 (2019). https://doi.org/10.1109/ACCESS.2019.2936435

34. L. Feng, W. Li, Y. Lin, L. Zhu, S. Guo, Z. Zhen, Joint computation offloading and URLLC resource allocation for collaborative MEC assisted cellular-V2X networks. IEEE Access **8**, 24914–24926 (2020). https://doi.org/10.1109/ACCESS.2020.2970750
35. Y. Sun, S. Zhou, J. Xu, EMM: energy-aware mobility management for mobile edge computing in ultra dense networks. IEEE J. Sel. Areas Commun. **35**(11), 2637–2646 (2017). https://doi.org/10.1109/JSAC.2017.2760160
36. J.L.D. Neto, S.-Y. Yu, D.F. Macedo, J.M.S. Nogueira, R. Langar, S. Secci, ULOOF: a user level online offloading framework for mobile edge computing. IEEE Trans. Mob. Comput. **17**(11), 2660–2674 (2018). https://doi.org/10.1109/TMC.2018.2815015
37. T. He, H. Khamfroush, S. Wang, T. La Porta, S. Stein, It's hard to share: joint service placement and request scheduling in edge clouds with sharable and non-sharable resources, in *2018 IEEE 38th International Conference on Distributed Computing Systems (ICDCS)* (2018), pp. 365–375. https://doi.org/10.1109/ICDCS.2018.00044
38. Y. Dai, D. Xu, S. Maharjan, Y. Zhang, Joint computation offloading and user association in multi-task mobile edge computing. IEEE Trans. Veh. Technol. **67**(12), 12313–12325 (2018). https://doi.org/10.1109/TVT.2018.2876804
39. X. Chen, W. Li, S. Lu, Z. Zhou, X. Fu, Efficient resource allocation for on-demand mobile-edge cloud computing. IEEE Trans. Veh. Technol. **67**(9), 8769–8780 (2018). https://doi.org/10.1109/TVT.2018.2846232
40. X. Ma, S. Zhang, W. Li, P. Zhang, C. Lin, X. Shen, Cost-efficient workload scheduling in cloud assisted mobile edge computing, in *2017 IEEE/ACM 25th International Symposium on Quality of Service (IWQoS)* (2017), pp. 1–10. https://doi.org/10.1109/IWQoS.2017.7969148
41. Y. Xiao, M. Krunz, QoE and power efficiency tradeoff for fog computing networks with fog node cooperation, in *IEEE INFOCOM 2017 - IEEE Conference on Computer Communications* (2017), pp. 1–9. https://doi.org/10.1109/INFOCOM.2017.8057196
42. F. Wang, J. Xu, S. Cui, Optimal energy allocation and task offloading policy for wireless powered mobile edge computing systems. IEEE Trans. Wirel. Commun. **19**(4), 2443–2459 (2020). https://doi.org/10.1109/TWC.2020.2964765
43. Y. Yang, C. Long, J. Wu, S. Peng, B. Li, D2D-enabled mobile-edge computation offloading for multiuser IoT network. IEEE Int. Things J. **8**(16), 12490–12504 (2021). https://doi.org/10.1109/JIOT.2021.3068722
44. Z. Zhang, W. Zhang, F. Tseng, Satellite mobile edge computing: improving QoS of high-speed satellite-terrestrial networks using edge computing techniques. IEEE Netw. **33**(1), 70–76 (2019). https://doi.org/10.1109/MNET.2018.1800172
45. G. Zhang, W. Zhang, Y. Cao, D. Li, L. Wang, Energy-delay tradeoff for dynamic offloading in mobile-edge computing system with energy harvesting devices. IEEE Trans. Ind. Inf. **14**(10), 4642–4655 (2018). https://doi.org/10.1109/TII.2018.2843365
46. B. Li, Z. Fei, J. Shen, X. Jiang, X. Zhong, Dynamic offloading for energy harvesting mobile edge computing: architecture, case studies, and future directions. IEEE Access **7**, 79877–79886 (2019). https://doi.org/10.1109/ACCESS.2019.2922362
47. F. Zhao, Y. Chen, Y. Zhang, Z. Liu, X. Chen, Dynamic offloading and resource scheduling for mobile-edge computing with energy harvesting devices. IEEE Trans. Netw. Serv. Manag. **18**(2), 2154–2165 (2021). https://doi.org/10.1109/TNSM.2021.3069993
48. F. Guo, H. Zhang, X. Li, H. Ji, V.C.M. Leung, Joint optimization of caching and association in energy-harvesting-powered small-cell networks. IEEE Trans. Veh. Technol. **67**(7), 6469–6480 (2018). https://doi.org/10.1109/TVT.2018.2805370
49. H. Wu, L. Chen, C. Shen, W. Wen, J. Xu, Online geographical load balancing for energy-harvesting mobile edge computing, in *2018 IEEE International Conference on Communications (ICC)* (2018), pp. 1–6. https://doi.org/10.1109/ICC.2018.8422299
50. L.P. Qian, Y. Wu, B. Ji, X.S. Shen, Optimal ADMM-based spectrum and power allocation for heterogeneous small-cell networks with hybrid energy supplies. IEEE Trans. Mob. Comput. **20**(2), 662–677 (2021). https://doi.org/10.1109/TMC.2019.2948014

51. G. Zhang, Y. Chen, Z. Shen, L. Wang, Distributed energy management for multiuser mobile-edge computing systems with energy harvesting devices and QoS constraints. IEEE Int. Things J. **6**(3), 4035–4048 (2019). https://doi.org/10.1109/JIOT.2018.2875909
52. C. You, K. Huang, H. Chae, B.-H. Kim, Energy-efficient resource allocation for mobile-edge computation offloading. IEEE Trans. Wirel. Commun. **16**(3), 1397–1411 (2017). https://doi.org/10.1109/TWC.2016.2633522
53. J. Zhang et al., Energy-latency tradeoff for energy-aware offloading in mobile edge computing networks. IEEE Int. Things J. **5**(4), 2633–2645 (2018). https://doi.org/10.1109/JIOT.2017.2786343
54. Y. Geng, Y. Yang, G. Cao, Energy-efficient computation offloading for multicore-based mobile devices, in *IEEE INFOCOM 2018 - IEEE Conference on Computer Communications* (2018), pp. 46–54. https://doi.org/10.1109/INFOCOM.2018.8485875
55. Y. Geng, G. Cao, Peer-assisted computation offloading in wireless networks. IEEE Trans. Wirel. Commun. **17**(7), 4565–4578 (2018). https://doi.org/10.1109/TWC.2018.2827369

Chapter 4
Deep Reinforcement Learning for Delay-Aware and Energy-Efficient Computation Offloading

4.1 System Model and Problem Formulation

4.1.1 System Model

Consider in the system there exists multiple end users $\mathcal{U}_t = \{u_1, u_2, \ldots, u_N\}$ at a given time slot t, where N is the total number of users. There are also multiple MEC servers $\mathcal{M}_t = \{m_1, m_2, \ldots, m_K\}$ that are in proximity of end devices, where K is the total number of MEC servers. In practice, MEC servers can be with different computing capacities and workloads, and the heterogeneity of MEC servers can be captured by the set of $\mathcal{M}_t = \{(Q_1, f_{max}^{(1)}), \ldots, (Q_k, f_{max}^{(k)}), \ldots, (Q_K, f_{max}^{(K)})\}$, where Q_k and $f_{max}^{(k)}$ denote the task queue and maximum CPU frequency of the kth server. At a given time slot, the end devices may generate different tasks and select to offload more than one tasks to the edge servers for processing or they may process the tasks locally. The set of tasks is denoted as $\Omega_t = \{\Omega_{0,0}, \Omega_{0,1}, \ldots, \Omega_{i,j}, \ldots, \}$, where $\Omega_{i,j}$ represents the jth task of user i and is defined as $\Omega_{i,j} = \{D_{i,j}, C_{i,j}, \Delta_{max}\}$; where $D_{i,j}, C_{i,j}, \Delta_{max}$ denote the data size, the requested CPU cycles, and the maximum tolerable time of the task (i.e., the required deadline), respectively. Suppose in the system there is a coordinator/control agent that can communicate with the MEC servers to coordinate the task offloading and resource allocation. Such a coordinator can be implemented on any of the local MEC or cloud servers. The control agent will collect the state information regarding the system, including the tasks profile and status of each MEC server, and run a learning model to output the optimal decisions for task offloading and task execution. The action that the agent can take includes the index of the MEC server for offloading a certain task, and the CPU frequency at MEC servers to execute the tasks. Then, following the instructions from the control agent, the tasks will be offloaded to the selected MEC servers, and the servers will process the tasks with the

Y. Chen et al., *Energy Efficient Computation Offloading in Mobile Edge Computing*, Wireless Networks, https://doi.org/10.1007/978-3-031-16822-2_4

Table 4.1 Notations and definitions

$u_n^{(t)}$	nth User at time slot t
$m_k^{(t)}$	kth MEC server at time slot t
$T_{i,j}^{(t)}$	jth task from ith user at the t time slot
$D_{i,j}$	Data size of task $\Omega_{i,j}^{(t)}$
$C_{i,j}$	Demanded CPU cycles to process $T_{i,j}^{(t)}$
Δ_{max}	Maximum tolerate time of task $\Omega_{i,j}^{(t)}$
Q_k	Tasks queue in MEC server m_k
$f_{max}^{(k)}$	Maximum frequency of MEC server m_k
δ_k^R	Remain running time of the current task running
$\delta_{i,j}^T$	Transfer time of task $\Omega_{i,j}$
$\delta_{i,j}^C$	Computing time of task $\Omega_{i,j}$
δ_k^Q	Waiting for tasks in the queue time
$\zeta_{i,j}$	Transfer speed of the connection

recommended CPU frequencies. The main notations and definitions in the system model are given in Table 4.1.

The objective is to maximize the number of tasks that are completed before the required deadlines while minimizing the energy consumption. A task $\Omega_{i,j}$ is considered to be successfully completed if the total service latency is less than its maximum tolerable time Δ_{max}; otherwise the task fails, formally,

$$F_{task} = \begin{cases} 1, & \text{if } \delta_k^R + \delta_k^Q + \delta_{i,j}^T + \delta_{i,j}^c \leq \Delta_{\max}\ ; \\ 0, & \text{otherwise.} \end{cases} \tag{4.1}$$

The total service latency for task $\delta_{i,j}$ includes the residual time δ_k^R of the current task running at the selected server, the transmission time to offload the task to the edge server $\delta_{i,j}^T$, the waiting time in the queue before service δ_k^Q, and the computing time $\delta_{i,j}^C$. The residual running time δ_k^R of the current task being executed at the target server can be obtained as the total computing time $\delta_{current}^C$ of the current task minus the start running time δ_{run}, namely,

$$\delta_k^R = \delta_{current}^C - \delta_{run}. \tag{4.2}$$

The waiting time in the queue δ_k^Q is actually the summation of the computing time for all the tasks before the current task in the queue. The computing time for a particular task can be obtained by dividing the required CPU cycles $C_{*,j}$ [1] by the recommended frequency $f_{*,j}$. Thus, the waiting time can be given by

$$\delta_k^Q = \sum_{j=0}^{M} \frac{C_{*,j}}{f_{*,j}}. \tag{4.3}$$

The transmission time $\delta_{i,j}^T$ can be given as below

$$\delta_{i,j}^{T} = \frac{D_{i,j}}{\zeta_{i,j}}, \tag{4.4}$$

where $D_{i,j}$ is the data size of the task and $\zeta_{i,j}$ is the transmission rate. The transmission rate $\zeta_{i,j}$ [2] can be given by

$$\zeta_{i,k} = \alpha_{i,l}^{(B)} log_2(1 + \frac{P_{i,k} h_{i,k} L_{i,k}}{N_0}), \tag{4.5}$$

where $\alpha_{i,l}^{(B)}$ is the bandwidth, $P_{i,l}$ is the transmission power, and $h_{i,l}$, $L_{i,l}$ are Rayleigh fading and path loss, respectively. Suppose that the task is offloaded to the kth, and the computing time $\delta_{i,j}^{C}$ can be calculated by

$$\delta_{i,j}^{C} = \frac{C_{i,j}}{f_{i,j}^{k}}, \tag{4.6}$$

where $C_{i,j}$ is the requested CPU cycles to compute task $\Omega_{i,j}$ and the recommended frequency $f_{i,j}^{k}$.

For the total energy consumption $E_{i,j}$, it is summation of the energy consumption for transmission denoted by $E_{i,j}^{T}$ and energy consumption caused by computation denoted by $E_{i,j}^{C}$, that is,

$$E_{i,j} = E_{i,j}^{T} + E_{i,j}^{C}. \tag{4.7}$$

Note that the energy consumption due to transmission is given by

$$E_{i,j}^{T} = \delta_{i,j}^{T} P_{i,j} = \frac{D_{i,j}}{\zeta_{i,j}} P_{i,j}. \tag{4.8}$$

The energy consumption for computation [3] can be given by

$$E_{i,j}^{C} = c(f_{i,j}^{k})^2 C_{i,j}, \tag{4.9}$$

where $c = 10^{-26}$ [4, 5] and $f_{i,j}^{k}$ is the frequency used to execute the task $\Omega_{i,j}$ at the kth edge server.

4.1.2 Problem Formulation

We focus on the long-term utility of the system. Specifically, we aim to maximize the number of tasks completed before the deadlines and minimize the energy consumption in the long run. To this end, we formulate a Markov Decision Process

(MDP) to maximize expected long-term rewards. Concretely, when one or more MEC servers are overloaded, we consider an episode is terminated. Each episode contains many time slots denoted by t, and the reward R_t depends on the action a_t taken by the agent when the current state is s_t. The action space and state space are denoted by A_t and S_t, respectively.

Ideally, we can distribute the tasks received at a time slot with a single action. However, the RL model can get diverged with explosion of action space when distributing multiple tasks from multiple users to multiple MEC servers. The search action space would grow exponentially as the number of users, tasks, or the number of MEC servers increases. It is challenging to guarantee that the agent can learn and the learning model can converge to an optimal solution with the explosion of action space. Let N_Ω be the number of received tasks, and K be the number of MEC servers. Although the action spaces equal to K^{N_Ω}, the action space is much larger than the exponential function of the number of tasks, because we need to consider the order of the tasks in each server. For example, we can enumerate the all possible actions when we have three MEC servers $\{m_1, m_2, m_3\}$ and three tasks arrived at the same time slot, and the action space can be computed by

$$\begin{aligned} A_s &= (0! \times 0! \times P_3^3) + (0! \times P_1^3 \times P_2^3) \\ &+ (0! \times P_2^3 \times 1!) + (0! \times P_3^3 \times 1!) \\ &+ (P_1^3 \times 0! \times P_2^2) + (P_1^3 \times P_1^2 \times 1!) \\ &+ (P_1^3 \times P_2^2 \times 0!) + (P_2^3 \times 0! \times 1!) \\ &+ (P_2^3 \times 1! \times 0!) + (P_3^3 \times 0! \times 0!) \\ &= 10 \times 3! = 60, \end{aligned} \tag{4.10}$$

where, $(0! \times 0! \times P_3^3)$ is the action space for offloading 0 task to m_1, 0 task to m_2, and 3 tasks to m_3; and $(0! \times P_1^3 \times P_2^3)$ is the action space for offloading 0 task to m_1, 1 task to m_2, and 2 tasks to m_3, and so on.

We can consider offloading identical tasks to different servers with different orders. We then can use permutation and combination function to derive the action space size as:

$$A_s = P_{N_\Omega}^{N_\Omega} \times C_{K-1}^{(N_\Omega+1)+(K-2)} = N_\Omega! \times C_{K-1}^{N_\Omega+K-1} \tag{4.11}$$

In such a case, the model also recommends frequencies to execute the tasks. Let f_k denote the action size of recommended frequencies; and then the action space can be given by

$$A_s = N_\Omega! \times C_{K-1}^{N_\Omega+K-1} \times f_k. \tag{4.12}$$

Moreover, if tasks are processed in parallel, an immense input size is needed for the learning model to receive a large number of tasks in peak-hours. However, in off-peak hours, the input feature for the learning model would be considerably sparse. As a result, the learning model can probably process matrices that are full of zeros which can waste the MEC resources most of the time.

To address the aforementioned challenges, we process the received tasks sequentially within a time slot. Suppose that the channel and other parameters of the MEC remain the same during time slot t, and the transition of states only depends on the actions taken by the control agent. Furthermore, the controller of the MEC network can collect profiles of the tasks in a queue, and the learning agent can take actions with respect to task queue within τ time step, where $\tau << t$. Similarly, the transition probability can be formed based on τ time step rather than time slot t (Eq.4.13). The state transition with time step τ satisfies the property of MDP, and the transition probability $p(s'|s, a)$ is given as follows:

$$p(s'|s, a) \doteq \Pr\{s_\tau = s'|s_{\tau-1} = s, a_{\tau-1} = a\}. \quad (4.13)$$

In what follows, we will present the details of the MDP in the MEC network environment with the time step τ within the given time slot t. The states at time slot t is given by

$$s_t = \{s_0, s_1, s_\tau, \ldots, s_\lambda | \sum_{\tau=0}^{\lambda} \tau << t\}, \quad (4.14)$$

where the λ is the number of tasks received at t. Note that $s_\tau = \{\Omega_\tau, M_\tau, \zeta_\tau\}$, where $m_k^{(\tau)}$ is the kth MEC server state in M_τ, and ζ_τ is the transmission rate matrix to ζ_τ servers.

Assume that at time slot t the number of tasks received by the MEC servers is no more than a threshold λ. The tasks with indices larger than λ will be processed in the next time slot. Moreover, the waiting time of the target server is updated when the tasks is added to the queues at the edge servers.

As stated in Sect. 4.2 each element of a general task queue at a MEC server is associated with the task information and recommended CPU frequency. Similarly, we can derive the task queues at the kth MEC server with tasks and the recommended frequency $f_r^{(k,\tau)}$ to process the task

$$Q_k^\tau = \{(\Omega_{1,1}^{(1,\tau)}, f_r^{(1,\tau)}), \ldots, (\Omega_{i,j}^{(k,\tau)}, f_r^{(k,\tau)}), \ldots\}, \quad (4.15)$$

where $\Omega_{i,j}$ is the jth task from the ith user and is to be distributed at time step τ.

Similarly, during time slot t the action that the agent can take within t is denoted as

$$a_t = \{a_0, a_1, a_\tau, \ldots, a_\lambda | \sum_{\tau=0}^{\lambda} \tau << t\}. \tag{4.16}$$

The actions that the agent can select include the index of edge servers for offloading tasks and the recommended CPU frequency at edge servers for executing the tasks. Therefore, the size of the action space is $K \times f_p$, where K is the number of MEC servers, and f_p is the resolution of discrete percents of the maximum CPU frequency of the edge server. A general action can be defined as $a_\tau = (k, f_r)$, where k is the target offloading server, and f_r is a recommended percentage of maximum frequency to execute the given task. The possible values of f_r can range from 0% to 100% .

For the reward function of the MDP framework, $R_t(s_t, a_t)$ is defined as the reward obtained in time slot t. Similar to the state and action, the reward at time slot t is a collection of rewards $R_\tau(s_\tau, a_\tau)$ within t, i.e.,

$$R_t(s_t, a_t) = \sum_{\tau=0}^{\lambda} R_\tau(s_\tau, a_\tau), \tag{4.17}$$

with $\sum_{\tau=0}^{\lambda} \tau << t$.

Then, the optimization problem is formulated as below:

$$\begin{aligned} \min_{a_\tau} \quad & \sum_{\tau=0}^{N} E_\tau^T + E_\tau^C \\ & = \sum_{\tau=0}^{N} \left[\frac{D_{i,j}}{\zeta_{i,j}} P_{i,j} + c(f_{i,j}^k)^2 C_{i,j} \right] \\ \text{s.t.} \quad & \delta_k^R + \delta_k^Q + \delta_{i,j}^T + \delta_{i,j}^c \leq \Delta_{\max}, \\ & f_{i,j}^k \leq f_k^{max}, \\ & a_\tau = \{k, f_{i,j}^k\} \end{aligned} \tag{4.18}$$

where E_τ^T and E_τ^C are the energy consumption for data transmission and data computation, respectively; f_k^{max} is the CPU capacity of the kth MEC server. However, in such a formulation there are several limitations. First, the constraint $\delta_k^R + \delta_k^Q + \delta_\tau^T + \delta_\tau^c \leq \Delta_{\max}$ may not be always satisfied, and the control agent has to deal with the cases when no solution exists in the feasible areas. Second, it is not flexible to balance the energy consumption and service latency. Third, the computational cost grows exponentially with the increase in the variables and the problem scale. Therefore, we will design a flexible reward function that allows the DRL model to solve all the optimization objectives simultaneously in an end-to-end manner. $R_\tau(s_\tau, a_\tau)$ is the reward at the time step τ which includes the number of the completed tasks and energy cost. It can be defined as follows:

$$R_\tau(s_\tau, a_\tau) = (1-\eta)\beta_1 F_{task} - \eta\beta_2 \log_2(E_\tau) + C, \tag{4.19}$$

where β_1 and β_2 are the normalizing terms for the reward from completing tasks on time and the cost due to the energy consumption. $\eta \in [0, 1]$ is a weight which can be used to balance those two objectives, i.e., the number of tasks completed before deadlines and the energy consumption. C is a positive constant which increases the accumulative rewards with the number of time slots. The accumulative reward due to C is equal to the number of time slots times C. With this constant, the agent will find the best policy to prevent the severs from being overloaded so that it can prolong the episodes and maintain the MEC network stability. Additionally, the energy cost is scaled with logarithm because it is approximately proportional to the square of the MEC CPU frequencies and considerably fluctuates.

Although the objective is to maximize expected long-term rewards, the learning agent can only have immediate reward from the current time step, while the rewards in future time steps are unknown. To evaluate the current action on the long-term reward, we utilize both the immediate reward and the expected rewards of the future which are estimated with learned policies. Specifically, the current action is evaluated by long-term return:

$$G_\tau \doteq R_\tau + \gamma R_{\tau+1} + \gamma^2 R_{\tau+2} + \cdots = \sum_{k=0}^{\infty} \gamma^k R_{\tau+k}, \tag{4.20}$$

which has the immediate reward and the discounted future rewards. $0 < \gamma < 1$ is called the discount factor. The immediate reward can be feedback from the environment, and the expected future rewards are computed with a policy π from the trained model. A policy π is a set of actions that the agent follows to interact with the environment.

Therefore, this work aims to develop a learning approach to find optimal policies π^* such that action-value function is maximized:

$$Q^*(s, a) = \max_\pi \mathbb{E}\left[R_\tau + \gamma G_{\tau+1} | s_\tau = s, a_\tau = a, \pi\right], \tag{4.21}$$

where R_τ is the immediate reward, and $G_{\tau+1}$ is the expected future reward discounted by γ.

4.2 Proposed DRL Method

In this section, we present the proposed DRL method to maximize the number of completed tasks and minimize the energy consumption by dynamically selecting the MEC servers for offloading the tasks and the allocating the computational frequency at the edge servers to process the tasks. Specifically, the proposed DRL model can learn and generate optimal policies that maximize the long-term reward defined in the previous section. By inputting observed data from the MEC network, the DRL model generates control parameters to maximize the number of completed tasks and minimize energy consumption. In the following, we will present the data preprocessing, the DRL model, and the training process, respectively.

4.2.1 Data Prepossessing

The data is considerably complicated and noisy, which increases the training efforts for the learning model. Data preprocessing is a critical to the proposed DRL method. The raw features, including time-varying channel condition, server states, and users' tasks. Moreover, since the ranges of raw features are significantly different, the proposed models might ignore the essential features. Finally, the dimension of input features is considerably high because the features include the channel distributions into the states, and the number of channels is increasing with the number of users and MEC servers. Without appropriate data preprocessing, it is possible that the DRL agent overlooks the essential features when working on the raw data. As a result, the agent converge too slowly for optimal solutions or converge to non-optimal solutions.

To deal with the above issues, we adopt normalization methods to re-scale and concatenate the features in a desirable format. The features containing hierarchical components are stored in a tree-like data structure. For instance, we have a root node, and a branch which represents features from the MEC servers. Furthermore, the branch has three sub-branches including MEC servers' state M_τ, transmission rate matrix ζ_τ, and the queue tasks Ω_τ, and each of them contains some leaf-level nodes. Therefore, we have to normalize the leaf-level sub-components and concatenate them together. First, the Frobenius norm (Eq. 4.22) is computed for all the leaf-level components of the feature $\mathbb{R}^{m\times n}$ [6]. Then, the normalization of the matrix $\mathbb{R}_N^{m\times n}$ is computed by dividing the normal in an element-wise manner, i.e.,

$$\|A\|_F = \sqrt{\sum_{i=1}^{m}\sum_{j=1}^{n} |a_{ij}|^2}, \tag{4.22}$$

where $a_{ij} \in \mathbb{R}^{m\times n}$ with

$$\mathbb{R}_N^{m\times n} = \frac{\mathbb{R}^{m\times n}}{\|A\|_F}. \tag{4.23}$$

Finally, all the normalized sub-features are concatenated as a single feature, which can feed to the learning model.

4.2.2 DRL Model

In this part, we propose a DRL model to address the joint optimization problem formulated in the previous section. The MEC network is regarded as the RL environment, and the proposed DRL model is considered as a learning agent that can interact with the MEC network and learn from the experience. Due to the complexity of the MEC environment, it is impossible that the states and transitions are fully observable to the agent. For simplicity, we consider the MEC network environment as a MDP with internal transition probability $P(r, s\prime|s, a)$, which is unknown to the DRL agent. Further, we propose a model-free DRL model to learn from the environment without knowing its internal transition. The DRL agent is expected to generate robust policies that can maximize long-term accumulated rewards.

4.2.2.1 Reinforcement Learning Framework

RL is a method that allows a learning agent to learn by interacting and exploring the unknown environment. Unlike standard machine learning, the RL models can learn from the sequential and evaluative feedback from the environment.

The RL agent needs to balance exploitation and exploration to learn from the unknown MEC environment. Exploitation is to use the learned knowledge by greedily exploring the search space according to the Q-value, namely,

$$a = \underset{a'}{\text{argmax}}\, Q\left(s, a'; w\right), \tag{4.24}$$

where w is the parameters matrix. On the other hand, exploration allows the learning agent to acquire new knowledge about the MEC environment by taking actions randomly. In this study, the $\epsilon - greedy$ method is adopted to balance exploitation and exploration, which means the model selects actions with a greedy algorithm with probability $1 - \epsilon$ and randomly selects actions with probability ϵ. Initially, the agent has no knowledge of the MEC network environment, and it takes more random actions in the early episodes to explore the environment. As the agent gradually acquires enough knowledge about the environment, the agent starts to exploit the learned knowledge to generate optimal policies. Therefore, ϵ is designed to decrease over the episodes.

Moreover, the goal of the RL agent is to learn the optimal policy π^* by finding the optimal action-values $Q^*(s, a)$ that maximizes the long-term accumulative rewards. Action-value $Q(s, a)$ is generated by taking action a state s, and then follows with the policy π. The optimal action-value $Q^*(s, a)$ is the maximum value of all possible values of $Q(s, a)$, i.e.,

$$Q^*(s, a) = \max_{\pi} \mathbb{E}\left[R_\tau | s_\tau = s, a_\tau = a, \pi\right]. \tag{4.25}$$

The policy that can optimize the action-value $Q^*(s, a)$ is the optimal policy π^* and satisfies the Bellman equation

$$Q(s, a) = \mathbb{E}_{s'}\left[R + \gamma \max_{a'} Q\left(s', a'\right) | s, a\right]. \tag{4.26}$$

The main idea of deriving optimal action-value (e.g., Q-value) is to take action a' from all possible actions for the next step of $Q(s, a)$ that maximizes $R + Q(s', a')$, and repeat it over all the states to generate optimal policies. Theoretically, we can derive the optimal action-value by updating the Bellman equation iteratively. $Q(s, a)$ value will be improving over iterations, and the $Q_\tau \rightarrow Q^*$ as $\tau \rightarrow \infty$, as given by,

$$Q_{\tau+1}(s, a) = \mathbb{E}_{s'}\left[R + \gamma \max_{a'} Q_\tau\left(s', a'\right) | s, a\right], \tag{4.27}$$

where Q_τ is the Q-value at step τ, and Q^* is the optimal value-function.

But it is impractical to find an optimal Q-value by iterating over ∞ times. Consequently, the classical RL models can diverge from the optimal policy in a vast or continuous search space because it is nearly impossible to explore all of the search space. Fortunately, we can reduce the search space by approximate functions, such as linear and non-linear functions. A deep neural network can be considered as a non-linear function that can approximate various complex states, $Q(s, a; w) \approx Q^*(s, a)$.

4.2.2.2 Deep Reinforcement Learning Model

Due to the high complexity and the continuous states, it is impossible to store all the state-action value pairs in a Q-table to search the optimal policies. Although we can consider discretization of the continuous space to make the continuous space into discrete space, it is also challenging to balance the resolution of discrete space. On the one hand, the low-resolution discretization can compromise the accuracy of the representation of the features. On the other hand, a high-resolution discrete space would result in a vast search space that dramatically increases the search time and complexity. Furthermore, enormous search space hinders the model from convergence and finding the optimal policies. Therefore, we adopt a deep neural network as an approximator to represent the search space. Specifically, the deep neural network represents input states, and the model computes probabilities of all possible actions $P_A = \{p_1, \ldots, p_k\}$ at one time. The agent selects the actions based on the probabilities and interactions with the environment.

The offloading system is as shown in Fig. 4.1. First, a coordinator deployed in the MEC network collects the state information of the system and provides an interface to the DRL agent. Specifically, the coordinator first collects offloading tasks' profiles and puts them into the queue in the MEC environment (MEC ENV).

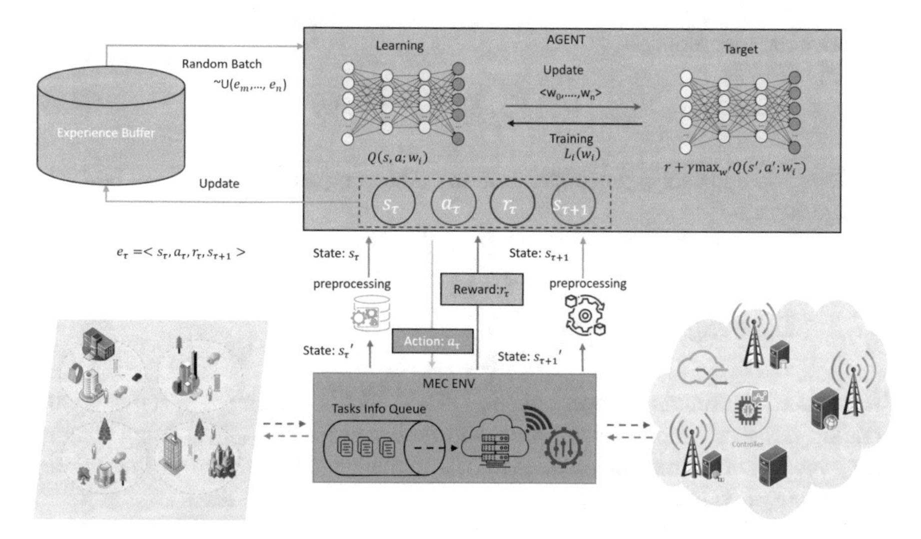

Fig. 4.1 Offloading system

Second, the DRL agent takes action based on the state information observed from the MEC environment. Third, the coordinator executes the action by offloading tasks to the selected servers, and then the MEC servers process the tasks using the CPU frequencies recommended by the DRL agent. Fourth, the DRL agent stores the data (state, action, reward, and next state) into the experience replay buffer for training the DRL model. Fifth, the DRL agent randomly draws sample data from the experience replay buffer to train the neural network by minimizing the loss function defined by a Mean Square Error (MSE). Finally, the target network is updated after every $\mathcal{N}$ episodes. The training steps can entirely separate from the above steps, which means we can run the training process with the above steps simultaneously.

In the proposed DRL model, the approximate function is a neural network (also known as Q-Network) with parameters w. Incorporating deep neural networks with RL is considerably difficult to train, as the unknown fluctuating feedback from the dynamic environment. For mitigating the oscillation and preventing divergence during training, the deep Q-network method introduces fixation methods. Specifically, the approximator neural network has a copy with fixed parameters w^-, also known as the target network, where the weights keeps unchanged in a certain number of episodes. The parameters w of the other neural network called the primary network (also known as local learning network) keeps learning from the data in the memory buffer, and its weights are copied to the target network after every $\mathcal{N}$ episodes. The parameters w_i are updated to minimize the loss function, which is the MSE between current action-value with $Q(s', a'; w_\tau)$ and optimal $Q^*(s', a')$, which can be substituted with fixation term $\bar{F}$

$$\bar{F} = r + \gamma \max_{a'} Q^* \left(s', a'; w_\tau^-\right), \tag{4.28}$$

to derive the loss function

$$L_\tau(w_\tau) = \mathbb{E}_{s,a,r}\left[\left(\mathbb{E}_s[\bar{F}|s,a] - Q\left(s,a;w_\tau\right)\right)^2\right]. \tag{4.29}$$

where w_τ^- is updated in previous iterations. The complete algorithm can be found in Algorithm 3.

4.2.3 Training

In this part, the training process of the DRL model is presented. Specifically, the training has three components: initialization and preparation, generating training data, and learning from the data. Also, the techniques introduced in the training process are provided in details.

Algorithm 3 DQN-Learning for MEC

1: **Input:** $epoch\_no$, ϵ_{start}, ϵ_{end}.
2: **Output:** $loss$, $gradients$.
3: //1. Initialization:
4: Initialize replay memory D with capacity N;
5: Initialize action-value Q with random weights w;
6: Initialize target action-value $\hat{Q}$ weights $w^- \leftarrow w$;
7: Initialize scores with window size;
8: $\epsilon \leftarrow \epsilon_{start}$;
9: **for** $episode \leftarrow 1$ to M **do**
10: Initialize input raw data x_1;
11: Prepossess initial state: $S \leftarrow \phi(< x_1 >)$;
12: **for** time step: $\tau \leftarrow 1$ to T_{max} **do**
13: // 2. Generate training data:
14: Select action A from state S using: $\pi \leftarrow \epsilon - Greedy(\hat{Q}(S, A, w))$;
15: Take action A, Observe reward R and get next input $s_{\tau+1}$;
16: Prepossessing next state: $S' \leftarrow \phi(s_{\tau+1})$;
17: Store experience tuple (S, A, R, S') in replay memory D;
18: $S' \leftarrow S$;
19: // 3. Learning:
20: Obtain random mini-batch of (s_j, a_j, r_j, s_{j+1}) from D;
21: **if** episode terminate at step $j+1$ **then**
22: Set target $\bar{F}_j \leftarrow r_j$;
23: **else**
24: Set target $\bar{F}_j \leftarrow r_j + \gamma \max_{a'} \hat{Q}\left(s_j, a, w^-\right)$;
25: **end if**
26: Update: $w \leftarrow w + \alpha \nabla_{w_j} L\left(w_j\right)$ with $Adam$;
27: Every N steps, update: $w^- \leftarrow w$;
28: **end for**
29: $\epsilon \leftarrow max(\epsilon_{end}, \epsilon * decay)$;
30: Store score for current episode;
31: **end for**

4.2.3.1 Initialization

As shown in Algorithm3, the algorithm initializes the experience replay memory buffer, the exploration proportion ϵ, and two neural networks including the learning and target networks. The replay buffer stores experience of the DRL agent when interacting with the MEC network environment. A learning network is initialized with random weights and replicated to the target network. Then, the algorithm iteratively generates data and trains the DRL model over the episodes. The episode ends when the time step is larger than or equal to the threshold T_{max}, or the environment returns a finish flag, which indicates at least one of the MEC servers is overloaded. For a regular RL framework, the MEC network environment resets whenever the learning algorithm starts a new episode. The reset action cleans all the tasks and other settings in all of the MEC servers.

4.2.3.2 Exploration and Data Acquisition

The DRL agent keep interacting with the MEC environment to generate the training dataset. In particular, as mentioned earlier, the DRL agent acquires knowledge through a $\epsilon - greedy$ method. The agent randomly explores the environment and generates the greedy actions with a probability of ϵ, or it selects other actions with a probability of $1 - \epsilon$. Each interaction generates a tuple containing the current state s_τ, action a_τ, reward r_τ and $s_{\tau+1}$. Furthermore, the generated data is kept in the experience buffer for learning purposes. Additionally, the data generation module and the learning module do not depend on each other; therefore, these two modules do not need to follow each other step by step. For instance, for the learning module, the model can have several runs of the data generation or a single run; and they can run separately and simultaneously.

4.2.3.3 Replay Experience Buffer

The sequences of the experience-tuples could be highly correlated when the agent interacts with the MEC environment if the data is fed into the model sequentially. The classical Q-learning methods learning from sequentially ordered data cause risks of being swayed due to correlation among data. To prevent action values from oscillating or diverging, we adopt experiences replay method to draw training data uniform randomly from the experiences buffer. With this approach, instead of learning data timely as interacting with the environment, the agent collects experiences tuples $< s, a, r, s' >$ into the experiences buffer. The experience buffer is a queue with a fixed size, and the generated data will be added into the queue over time. The oldest data will be deleted to make available room for new data when the buffer is full. With experiences replay and random samples, the actual loss function can be given as

$$L_\tau(w_\tau) = \mathbb{E}_{(s,a,r,s')\sim U(D)}\Big[\Big(r + \gamma \max_{a'} Q(s', a'; w_\tau^-) - Q(s, a; w_\tau)\Big)^2\Big]. \tag{4.30}$$

The experience buffer can enhance the training process in the following ways. First, the correlation of the sequential order can be decoupled by sampling the training data from the experience buffer instead of feeding the sample one by one. Second, the experience replay allows the agent to learn more from individual entry multiple times. More importantly, the experience replay can recall rare occurrences to prevent the model from overfitting due to bias of training sample distribution. Lastly, it can mitigate the oscillation or divergence caused by outlier training samples by using batch samples. The model is allowed to sample multiple data samples to leverage the batch normalization to reduce the swaying.

4.2.3.4 Learning

The DRL learning process is slightly different from training a conventional deep learning model. In the forward propagation, the model randomly draws a batch of samples from the experience buffer and feeds them to both the learning and target networks. Then, the loss (Eq. 4.29) is calculated based on the errors between the rewards from the learning and target networks. Further, using backpropagation, the parameters of the local learning network are updated, which is similar to training a regular neural network. However, unlike the loss function of the standard neural network which is computed as the error between the outputs and labels, the loss functions of DRL is computed by the difference between the outputs from the learning network and the target network because the DRL agent learns from evaluative feedback rather than true labeled data. To recap, the loss is computed in the forward propagation, and then the parameters w are adjusted w with learning rate α times partial derivative loss function with respect to w, as follows:

$$w \leftarrow Adam(w, \alpha \nabla_{w_\tau} L(w_\tau)). \tag{4.31}$$

To simplify the partial derivative of loss function, we first can derive the loss function as

$$L_\tau(w_\tau) = \mathbb{E}_{s,a,r,s'}\Big[\big(\bar{F} - Q(s, a; w_\tau)\big)^2\Big] + \mathbb{E}_{s,a,r}\Big[\mathbb{V}_s[\bar{F}]\Big]. \tag{4.32}$$

Since the last term $\mathbb{E}_{s,a,r}\Big[\mathbb{V}_s[\bar{F}]\Big]$ of the loss function does not depend on learning network parameters w, it can be neglected when computing the partial derivative with respect to w. In other words, the partial derivative of loss function is derived as:

$$\begin{aligned}\nabla_{w_\tau} L(w_\tau) &= \nabla_{w_\tau} \mathbb{E}_{s,a,r,s'}\left[\left(\bar{F} - Q(s,a;w_\tau)\right)^2\right] \\ &\quad + \nabla_{w_\tau} \mathbb{E}_{s,a,r}\left[\mathbb{V}_s[\bar{F}]\right] \\ &= \nabla_{w_\tau} \mathbb{E}_{s,a,r,s'}\left[\left(\bar{F} - Q(s,a;w_\tau)\right)^2\right];\end{aligned} \tag{4.33}$$

further, the gradient of $L(w_\tau)$ is derived by using the chain rule and substituting $\bar{F}$ with $\bar{F} = r + \gamma \max_{a'} Q^*\left(s', a'; w_\tau^-\right)$. Therefore, the gradient of $L(w_\tau)$ is:

$$\begin{aligned}\nabla_{w_\tau} L(w_\tau) = \mathbb{E}_{s,a,r,s'}\Big[\Big(r + \gamma \max_{a'} Q\left(s,', a'; w_\tau^-\right) \\ -Q(s,a;w_\tau)\Big)\nabla_{w_\tau} Q(s,a;w_\tau)\Big].\end{aligned} \tag{4.34}$$

where $w_\tau^- = w_{\tau-1}$. Additionally, Adam [7] optimization function is adopted when updating the parameters. Finally, the updated parameters of learning network w are copied to the target network every N episodes to overwrite the w^-.

4.2.3.5 Reward Clipping

For convergence and generating the optimal policies smoothly, both the rewards and the loss errors can be clipped. Due to the complexity and uncertainty of the MEC network, the rewards obtained can be significantly different even with a small change of the feature. Also, some features, such as channel distributions, may distort the DRL model back and forth as the features have a wide range and high variance of the training samples. Consequently, the DRL model can be slow to converge, or it may never converge to the optimal polices. The clip is a straightforward but practical technique that can mitigate those issues. With clips, any element in $\mathbb{R}^{m \times n}$ less than min would be replaced with min, and greater than max would be replaced with max

$$\mathbb{R}^{m \times n}_{clipped} = clip(\mathbb{R}^{m \times n}, min, max). \tag{4.35}$$

4.3 Performance Evaluation

In this section, we present the simulation results to evaluate the performance of the DRL solution. In the simulations, Python is adopted as the programming language and processes are used to simulate the entities in the MEC system, including end

Table 4.2 Parameter setting

Signal to noise ratio (dB)	100
Task data size (bits)	$[2 \times 10^5, 2 \times 10^7]$
Task computing size (cycles)	$[8 \times 10^6, 1 \times 10^7]$
Server max frequency (Hz)	$[2 \times 10^9, 8 \times 10^9]$
Number of online user	[10, 1000]
Batch size	256
Learning rate α	5×10^{-4}
Discount Factor γ	0.9

users, MEC servers, the control agent, and the DRL agent. Additionally, PyTorch[1] and NumPy[2] are used to build a DRL model. We run the simulation with various parameter settings to compare the results and verify the proposed model.

As shown in Fig. 4.1, the simulation system also contains two major parts: the DRL agent and the MEC network environment (MEC ENV). The MEC ENV is made up of three components: the users, MEC servers, and the coordinator. In addition, the MEC network also maintains various states, such as the channel information and transmission rate. Some of the critical simulation parameters are summarized in Table 4.2. First, each user can create multiple tasks by waiting a random time period (around 0.001 seconds). Second, the simulator generates a set of MEC servers based on the parameter settings, including minimum and maximum frequencies, sizes of task queues, and overload thresholds. The MEC servers also maintain status information, process the tasks, and compute the rewards. The reward depends on the number of tasks that are completed before their deadlines and the energy consumption. Specifically, a MEC server executes the task using the recommended CPU frequency when it receives a task, and then the reward can be computed accordingly. The MEC is in the idle status with minimum frequency when the task queue is empty.

The DRL agent maintains the replay buffer and two neural networks, i.e., the local learning network and the target network. The DRL agent interacts with the MEC network through the coordinator, and then learns and generates actions and policies for the MEC network. In this simulation, the neural networks have five hidden layers, where the number of the neurons from layer one to layer five are 256, 512, 512, 512 and 256. Other parameters are given in Table 4.2. Each neural network has an input layer and an output layer, and the numbers of neurons are equal to the sizes of state space and action space, respectively.

Figure 4.2 shows the performance of proposed end-to-end DRL (E2E_DRL) model and existing DRL models. Note that the legends *E2E_DRL:10 users* and *E2E_DRL:1000 users* denote the proposed E2E_DRL model with 10 and 1000 users; *DRL:10 users* and *DRL:1000 users*, denote the existing DRL models with 10 and 1000 users. It can be seen that the models converge at around 500 episodes

[1] https://pytorch.org.

[2] https://numpy.org.

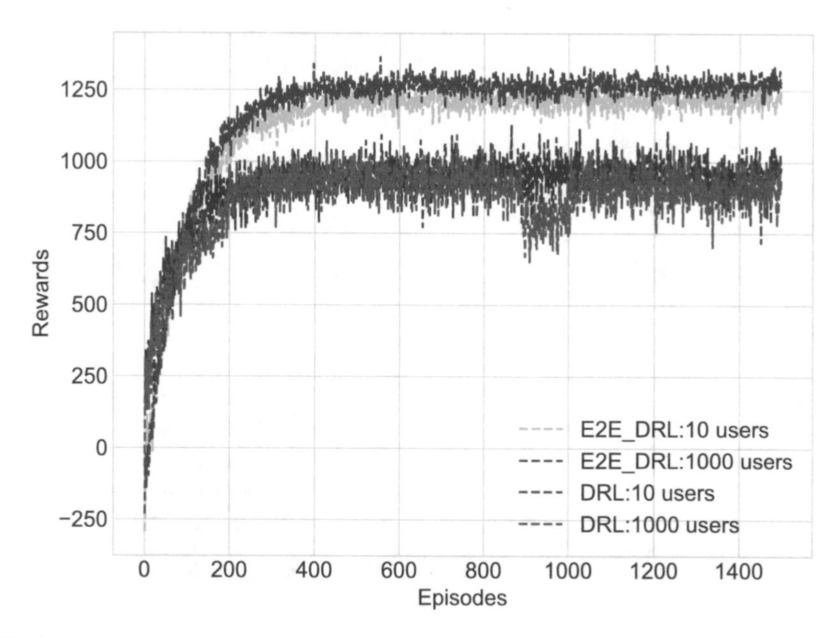

Fig. 4.2 Learning curve

and the proposed DRL model can obtain higher rewards than existing DRL models. The learning curves sway along with the episode as the generated data has many random factors, e.g., the uncertainty of tasks and the dynamics in MEC network environment. Compared with the proposed DRL model, the existing DRL models have larger variances and relatively lower rewards. That is because they only learn part of the decisions and the rest is dealt with by conventional optimization techniques. Additionally, optimization methods like CVX are designed for one-step optimization, rather than maximizing the expected long-term rewards. In contrast, the proposed DRL method learns all the actions to maximize expected long-term rewards.

Figure 4.3 compares the proposed model with some benchmarks including the greedy algorithm and other DRL models. From this figure, it can be seen that the proposed DRL algorithm outperforms the other models significantly. In addition, the DRL models can get more rewards when the number of users increases. That is because DRL models can learn better from more data, while the greedy algorithm is overwhelmed and performs poorly, with more users. Moreover, the proposed DRL model performing in an end-to-end manner, has more freedom to choose the actions than existing DRL. Note that the rewards fall at about the 850th to 900th episode when the DRL model serves 1000 users because the model takes some random actions to explore the environment.

For further comparison, we compare the proposed DRL model with the benchmarks in terms of the number of tasks completed and energy consumption, and the results are shown in Figs. 4.4 and 4.5, respectively. It can be seen that the

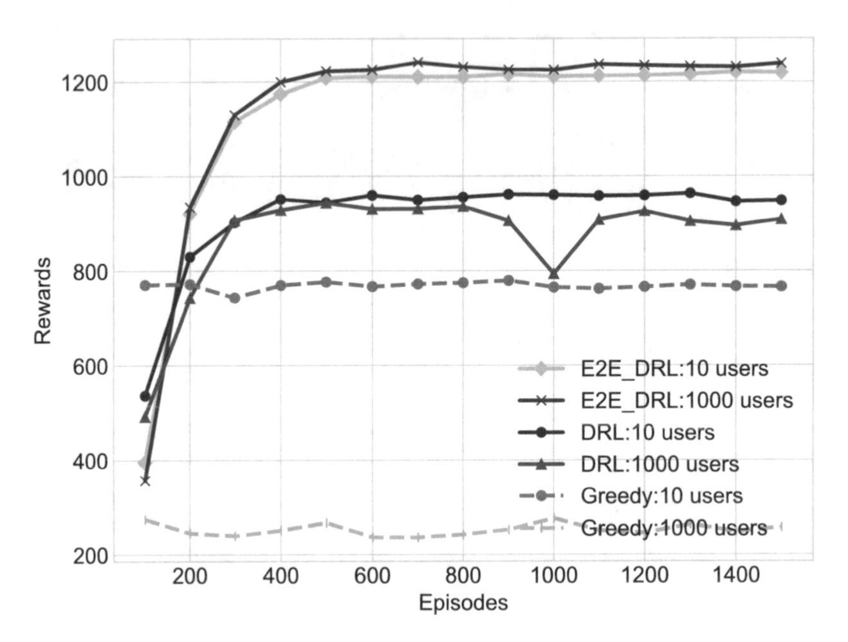

Fig. 4.3 Reward comparison

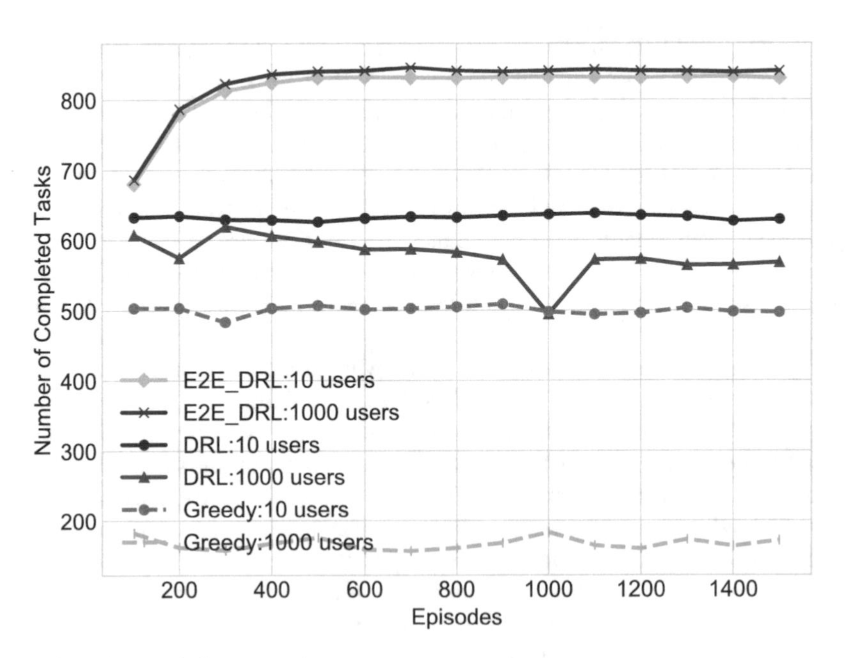

Fig. 4.4 Tasks completion comparison

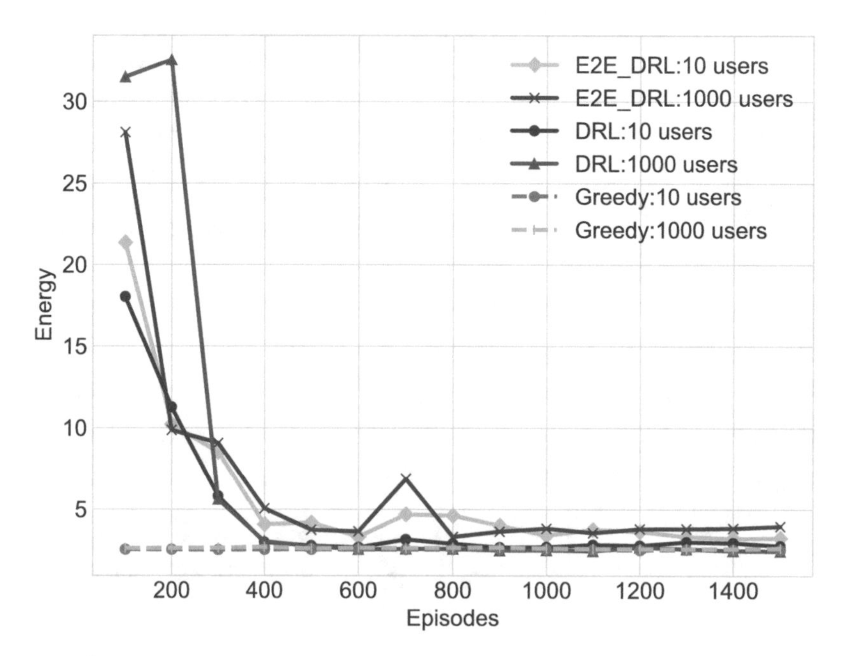

Fig. 4.5 Average energy consumption

proposed method outperforms the existing DRL models and the greedy algorithm. The performance of the proposed DRL models improves until convergence to the optimal policies because the DRL models keep learning from the historical data stored in the replay buffer. On the contrary, the greedy algorithm remains almost the same. Similarly, the DRL models consume more energy at the beginning episodes because the model takes action with $\epsilon - greedy$ and randomly initializes parameters in the early episodes. The proposed DRL model can save more energy because it learns from the acquired data and decreases random actions. Although the proposed DRL model is designed to maximize the long-term accumulated rewards, it also learns to reduce the energy cost over time. The DRL model with CVX can significantly saves energy consumption but does not increase the number of tasks because the DRL model can only decide part of the decision variables. The computational cost grows exponentially as the number of parameters and their search range increases.

In the proposed DRL model, by adjusting the weights in the reward function, the network operator can balance the number of tasks completed successfully before the deadlines and the energy cost. Note that the randomness of the simulation caused the sharp peak at about from the 600th to 800th episode in Figs. 4.5 and 4.6. The reasons for that are as follows. The energy consumption is proportional to the square of the CPU frequencies. Moreover, the computational cycles of the

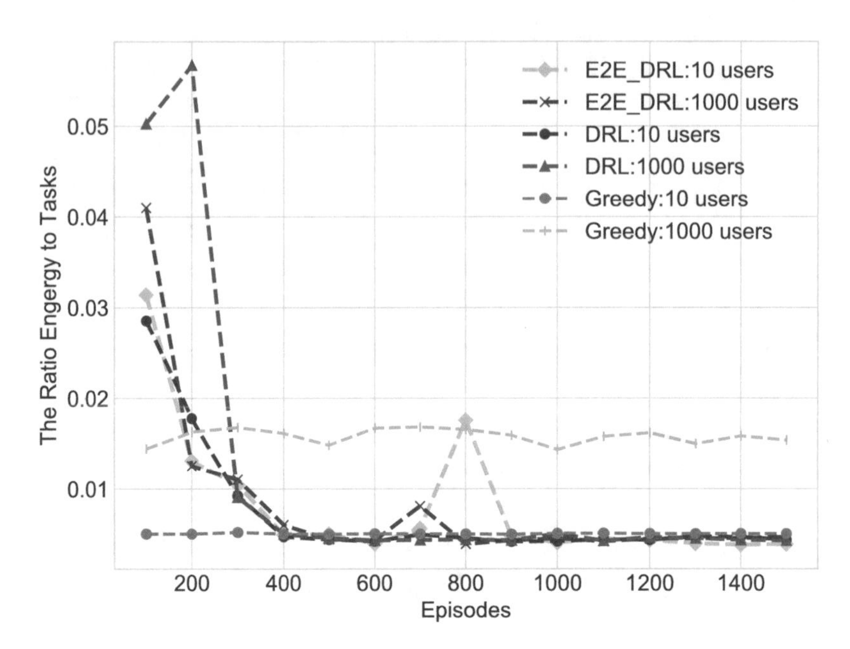

Fig. 4.6 The ratio of energy to tasks

tasks in the simulator could range from cycles 8×10^6 to 1×10^7. The learning algorithm incorporates ϵ-greedy, which means the algorithm would take some random actions to explore the environment, and those actions sometimes may cause considerable energy consumption. Therefore, the sharp point is raised by increasing computational cycles of tasks generated or the cost of exploitation.

Furthermore, Fig. 4.7 compares the rewards and convergence with different ϵ_{decay} values. The proposed DRL model uses ϵ_{decay} values to balance exploitation and exploration. The DRL model takes random actions with probability ϵ to explore, and $\epsilon = 1$ at the beginning of learning. It is desired that the model takes more random actions to explore the environment at the beginning and exploits the learned knowledge in the later episodes. Therefore, the value of ϵ is updated in each episode to decrease the exploration and increase exploitation, and it is updated in the following way: $\epsilon \leftarrow \epsilon_{decay} * \epsilon$. As seen from Fig. 4.7, the smaller ϵ_{decay} value, the faster the DRL model converges to the optimal policies. However, the models with a smaller ϵ_{decay} may not find the global optimal policies if they stop exploration too early. Therefore, the DRL models with a larger ϵ_{decay} values can achieve the models with small ϵ_{decay} values.

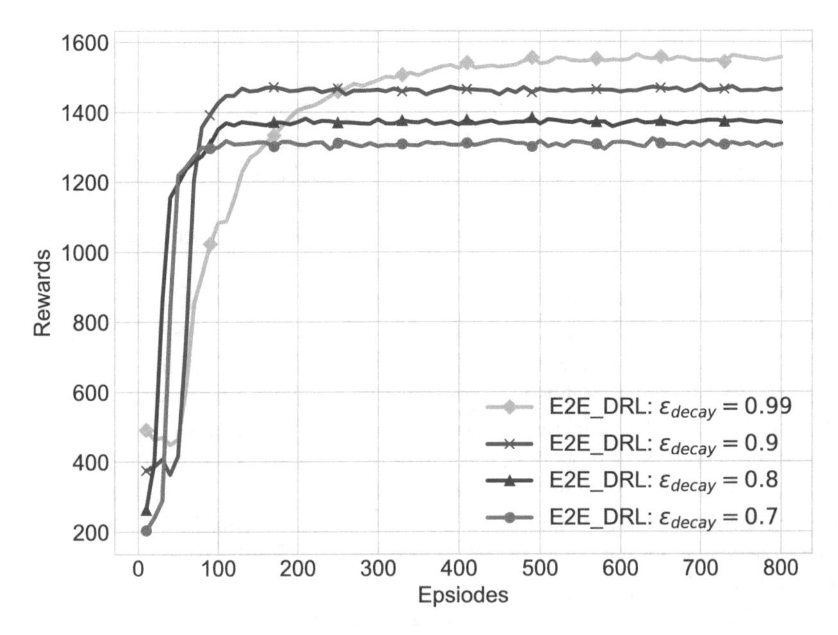

Fig. 4.7 ϵ_{decay} for end-to-end DRL model

4.4 Literature Review

As the enabling platform for a variety of promising applications, such as smart transportation and smart city [8–10], Internet of Things (IoT) expects to connect massive devices for data collection and data processing to support intelligent decision making and control. With the advantages of IoT, we have witnessed a rapid growth in IoT devices and applications. These IoT applications pose a high demand on storage and computing capacity, while the IoT devices are usually resource constrained. To efficiently process the data collected by the IoT devices, conventionally the data are sent to the remote cloud servers, where data processing and analysis are conducted for decision making. However, the cloud servers usually reside in the core networks, which is far away from the data source, i.e., IoT devices. Moving massive data in and out of the remote cloud servers can cause potential traffic congestion and long service latency.

To address the aforementioned issues, MEC emerges as a new paradigm, where computation and storage resources are deployed in proximity of the IoT devices [11, 12]. As a result, data generated by IoT devices can be offloaded to nearby MEC servers for processing, rather than sending data to the remote cloud servers [13]. As a result, the potential traffic congestion can be avoided, and the service latency can be significantly reduced. In addition, data security can be enhanced as the data Along with the benefits, MEC also encounters many challenges. Firstly, compared with cloud computing, MEC is usually with less capacity. Therefore, multiple MEC servers coordination is needed so that the tasks from IoT devices can be served efficiently. Secondly, some IoT services have stringent latency requirements, while service latency depends on many factors in offloading, such as transmission and computation power allocation. Thirdly, tasks with various service requirements are generated dynamically by IoT devices, and the MEC operating environment (e.g., channel conditions) changes over time, which requires the policy of the MEC system to be adjusted accordingly. Last but not least, energy consumption should also be considered, as IoT devices are energy constrained.

In the literature, various approaches based on optimization techniques are proposed to manage the MEC resources [14–16]. A joint computation offloading and system resource allocation for MEC is formulated as a mixed-integer non-linear programming in [17], which is then transformed into linear programming to reduce the complexity and tackle the challenges. Liang [18], employ linear-fractional programming (LFP), which is a generalization of linear programming (LP), and a greedy algorithm to optimize the offloading rate and energy consumption. In [19, 20], the tasks are divided into sub-task processes and processed on local devices or edge servers simultaneously based on optimization. Similarly, Gao [21] propose a method to find the optimal ratio of task partition into two sub-tasks for edge servers and local IoTs. He [22] propose an optimization approach to partition deep learning inference tasks and offload the partitions to edge servers for computing resource allocation. Those standard optimization methods are relatively straightforward to develop and address research problems in task offloading. However, the MEC system is usually very complex, and sometimes it is difficult to model in a mathematics format. Also, the optimization problem is mainly formulated based on a snapshot of the network, which has to be reformulated when the condition changes over time. Moreover, the majority of classical optimization methods require a large number of iterations, and they may find a local optimum rather than the global optimum.

Machine learning and deep learning [23–25] models can learn from historical labeled data and predict future computational offloading so that the system can make plausible decisions for MEC. Shen [26] surveyed machine learning methods for resource slicing and planning for the next-generation network. They have also summarized various machine learning and deep learning methods adopted in computational offloading and content offloading for MEC. Lyu [27] use stochastic gradient descent, a popular machine learning training method, to learn and partition data to offload spatially distributed edge servers; the authors argue that the proposed method can make optimal decisions for data partitioning with respect to time delay. Yang [28] proposed a statistical machine learning method to minimize the energy consumption for edge inference [29, 30], where deep learning inference tasks are processed on MEC. Ale [31] introduced a deep recurrent neural network to capture and predict the user requests so that they can make decisions for the content offloading and allocate resources based on the prediction. Summarily, the computation offloading problems can be formulated as a supervised classification problem and minimize the cost using deep learning [32]. The deep learning models can also adopt bandwidth allocation optimization and maximize the system utility [33]. Machine learning and deep learning methods can learn and predict the offloading decisions optimized with respect to exploiting resources and reducing the cost for the MEC. However, machine learning and deep learning require training data with labels, which require enormous effort to collect and label data. Moreover, it is challenging for humans to label data from optimization problems because it is difficult to take optimal actions in such a complex system.

Reinforcement Learning (RL) [34] enables the agent to learn the optimal policy through interacting with the MEC network environment, without relying on the labeled data for training. Therefore, RL holds great potential in MEC to facilitate model-free control without knowing the internal transition of the system. In [35], RL is utilized for MEC microservices coordination, and Q-learning is adopted for computation offloading in [36], and for optimizing remission rate and content delivery in [37]. Those standard RL methods leverage current and historical data for long-term decision-making, considering the current and the future rewards. Classical RL algorithm stores the learning results into a Q-table with tuples, consisting of states, action, and values. However, due to the Q-table which represents the policy, it is hard to be applied to ample search space. Therefore, RL methods might not be a desirable option for real-world MEC resources allocation and optimization due to the large size of the MEC states which is beyond Q-learning capacity.

Deep Reinforcement Learning (DRL) [38, 39] can help address the above issues in MEC, which integrates neural network into RL framework to approximate the Q values. In [40], a DRL based method is proposed to control computation offloading and minimize energy consumption for the Internet of Vehicles (IoV). Chen [41] propose a multiple-agent learning with Long Short-Term Memory (LSTM) to allocate resources for video stream in wireless networks. Similarly, an asynchronous advantage actor-critic (A3C) based model is proposed to render offloading Virtual Reality (VR) video streaming [42]. To improve the Quality-of-Service (QoS), a resource allocation approach is proposed in [43] to minimize the service latency. Similarly, a Q-Network learning approach is employed in [44] to maximize the number of successful transmissions. A DRL based collaborative MEC scheme is reported in [45] to minimize the service latency and energy consumption. However, the action space is relatively small, where the agent can only take two actions, i.e., offloading to MEC servers or executing the task locally. An online DRL based offloading control approach for MEC is proposed in [46]. Although both [45, 46] aim to optimize multiple objectives, DRL is only for partial of the problem and traditional optimization methods (e.g., linear programming) are employed to deal with the rest based on the outputs of the DRL. Although some issues can be addressed using classical optimization methods, given the output from DRL, the optimization still only focuses on the current time step parameters. Moreover, the DRL models are only trained to optimize the partial target, which cannot fully exploit its advantages to learn the overall best policy or even squeeze other targets. Therefore, it is would quite a challenge to address joint optimization problems by combining DRL and traditional optimization function to get optimal for all of the optimization targets. Different from them, in this chapter, an end-to-end

DRL approach is proposed to jointly determine the edge servers for offloading and allocate computing resources.

4.5 Summary

In this chapter, we presented a computation offloading problem in a dynamic MEC network, and presented an end-to-end DRL method to jointly optimize the edge server selection for offloading and the computing resource allocation, to maximize the number of tasks completed on time and minimize the energy consumption simultaneously. The presented DRL method can maximize long-term accumulated rewards instead of a one-time step. Moreover, it can make all the decisions without relying on other optimization functions to achieve joint optimization purposes. By applying the experience buffer replay and clip techniques, we can prevent the DRL model training process from suffering from oscillation and divergence.

References

1. J. Zhang, X. Hu, Z. Ning, E.C. Ngai, L. Zhou, J. Wei, J. Cheng, B. Hu, Energy-latency tradeoff for energy-aware offloading in mobile edge computing networks. IEEE Internet Things J. **5**(4), 2633–2645 (2018)
2. R. Bultitude, Measurement, characterization and modeling of indoor 800/900 Mhz radio channels for digital communications. IEEE Commun. Mag. **25**(6), 5–12 (1987)
3. Y. Mao, C. You, J. Zhang, K. Huang, K.B. Letaief, A survey on mobile edge computing: the communication perspective. IEEE Commun. Surv. Tuts. **19**(4), 2322–2358 (2017)
4. Y. Wang, M. Sheng, X. Wang, L. Wang, J. Li, Mobile-edge computing: partial computation offloading using dynamic voltage scaling. IEEE Trans. Commun. **64**(10), 4268–4282 (2016)
5. S. Guo, J. Liu, Y. Yang, B. Xiao, Z. Li, Energy-efficient dynamic computation offloading and cooperative task scheduling in mobile cloud computing. IEEE Trans. Mobile Comput. **18**(2), 319–333 (2019)
6. G.H. Golub, C.F.V. Loan, *Matrix Computation*, 4th edn. (The Johns Hopkins University Presss, Baltimore, 2013), p. 71
7. D.P. Kingma, J. Ba, Adam: a method for stochastic optimization. CoRR. vol. abs/1412.6980 (2014)
8. A. Al-Fuqaha, M. Guizani, M. Mohammadi, M. Aledhari, M. Ayyash, Internet of things: a survey on enabling technologies, protocols, and applications. IEEE Commun. Surv. Tuts. **17**(4), 2347–2376 (2015)
9. N. Lu, N. Zhang, N. Cheng, X. Shen, J.W. Mark, F. Bai, Vehicles meet infrastructure: toward capacity–cost tradeoffs for vehicular access networks. IEEE Trans. Intell. Transp. Syst. **14**(3), 1266–1277 (2013)
10. Q. Jiang, N. Zhang, J. Ni, J. Ma, X. Ma, K.K.R. Choo, Unified biometric privacy preserving three-factor authentication and key agreement for cloud-assisted autonomous vehicles. IEEE Trans. Veh. Technol. **69**(9), 9390–9401 (2020)
11. Y. Chen, N. Zhang, Y. Zhang, X. Chen, Dynamic computation offloading in edge computing for internet of things. IEEE Internet of Things J. **6**(3), 4242–4251 (2018)
12. A. Asheralieva, D. Niyato, Bayesian reinforcement learning and Bayesian deep learning for blockchains with mobile edge computing. IEEE Trans. Cognit. Commun. Netw. **7**, 319–335 (2021)

13. N. Zhang, N. Cheng, A.T. Gamage, K. Zhang, J.W. Mark, X. Shen, Cloud assisted hetnets toward 5G wireless networks. IEEE Commun. Mag. **53**(6), 59–65 (2015)
14. T.Q. Dinh, J. Tang, Q.D. La, T.Q.S. Quek, Offloading in mobile edge computing: task allocation and computational frequency scaling. IEEE Trans. Commun. **65**(8), 3571–3584 (2017)
15. P. Mach, Z. Becvar, Mobile edge computing: A survey on architecture and computation offloading. IEEE Commun. Surv. Tuts. **19**(3), 1628–1656 (2017)
16. M. Chen, Y. Hao, Task offloading for mobile edge computing in software defined ultra-dense network. IEEE J. Sel. Areas Commun. **36**(3), 587–597 (2018)
17. S. Bi, L. Huang, Y.A. Zhang, Joint optimization of service caching placement and computation offloading in mobile edge computing systems. IEEE Trans. Wireless Commun. **19**(7), 4947–4963 (2020)
18. Z. Liang, Y. Liu, T. Lok, K. Huang, Multiuser computation offloading and downloading for edge computing with virtualization. IEEE Trans. Wireless Commun. **18**(9), 4298–4311 (2019)
19. J. Liu, Q. Zhang, Adaptive task partitioning at local device or remote edge server for offloading in Mec, in *2020 IEEE Wireless Communications and Networking Conference (WCNC)* (2020), pp. 1–6
20. C. Fiandrino, N. Allio, D. Kliazovich, P. Giaccone, P. Bouvry, Profiling performance of application partitioning for wearable devices in mobile cloud and fog computing. IEEE Access **7**, 12156–12166 (2019)
21. J. Cao, L. Yang, J. Cao, Revisiting computation partitioning in future 5G-based edge computing environments. IEEE Internet Things J. **6**(2), 2427–2438 (2019)
22. W. He, S. Guo, S. Guo, X. Qiu, F. Qi, Joint DNN partition deployment and resource allocation for delay-sensitive deep learning inference in IoT. IEEE Internet Things J. **7**(10), 9241–9254 (2020)
23. Y. Lecun, Y. Bengio, G. Hinton, Deep learning. Nature **521**(7553), 436–444 (2015)
24. Y. LeCun, Y. Bengio, G. Hinton, Deep learning. Nature **521**(7553), 436–444 (2015). https://doi.org/10.1038/nature14539
25. J. Schmidhuber, Deep learning in neural networks: an overview. Neural Netw. **61**, 85–117 (2015)
26. X. Shen, J. Gao, W. Wu, K. Lyu, M. Li, W. Zhuang, X. Li, J. Rao, AI-assisted network-slicing based next-generation wireless networks. IEEE Open J. Veh. Technol. **1**, 45–66 (2020)
27. X. Lyu, C. Ren, W. Ni, H. Tian, R.P. Liu, E. Dutkiewicz, Optimal online data partitioning for geo-distributed machine learning in edge of wireless networks. IEEE J. Sel. Areas Commun. **37**(10), 2393–2406 (2019)
28. K. Yang, Y. Shi, W. Yu, Z. Ding, Energy-efficient processing and robust wireless cooperative transmission for edge inference. IEEE Internet Things J. **7**(10), 9456–9470 (2020)
29. X. Xu, Y. Ding, S.X. Hu, M. Niemier, J. Cong, Y. Hu, Y. Shi, Scaling for edge inference of deep neural networks. Nat. Electron. **1**(4), 216–222 (2018)
30. C. Xu, J. Ren, L. She, Y. Zhang, Z. Qin, K. Ren, Edgesanitizer: locally differentially private deep inference at the edge for mobile data analytics. IEEE Internet Things J. **6**(3), 5140–5151 (2019)
31. L. Ale, N. Zhang, H. Wu, D. Chen, T. Han, Online proactive caching in mobile edge computing using bidirectional deep recurrent neural network. IEEE Internet Things J. **6**(3), 5520–5530 (2019)
32. S. Yu, X. Wang, R. Langar, Computation offloading for mobile edge computing: a deep learning approach, in *2017 IEEE 28th Annual International Symposium on Personal, Indoor, and Mobile Radio Communications (PIMRC)*, (2017), pp. 1–6
33. H. Wu, Z. Zhang, C. Guan, K. Wolter, M. Xu, Collaborate edge and cloud computing with distributed deep learning for smart city internet of things. IEEE Internet Things J. **7**(9), 8099–8110 (2020)
34. R.S. Sutton, A.G. Barto, *Reinforcement Learning: An Introduction*, 2nd edn. (MIT Press, Cambridge, 2018)

35. S. Wang, Y. Guo, N. Zhang, P. Yang, A. Zhou, X.S. Shen, Delay-aware microservice coordination in mobile edge computing: a reinforcement learning approach. IEEE Trans. Mobile Comput. **20**, 939–951 (2021)
36. B. Dab, N. Aitsaadi, R. Langar, Q-learning algorithm for joint computation offloading and resource allocation in edge cloud, in *2019 IFIP/IEEE Symposium on Integrated Network and Service Management (IM)* (2019), pp. 45–52
37. Z. Su, M. Dai, Q. Xu, R. Li, S. Fu, Q-learning-based spectrum access for content delivery in mobile networks. IEEE Trans. Cognit. Commun. Netw. **6**(1), 35–47 (2020)
38. M. Riedmiller, Neural fitted Q iteration – first experiences with a data efficient neural reinforcement learning method, in *Machine Learning: ECML 2005*, ed. by J. Gama, R. Camacho, P.B. Brazdil, A.M. Jorge, L. Torgo (Springer, Berlin, 2005), pp. 317–328
39. V. Mnih, K. Kavukcuoglu, D. Silver, A.A. Rusu, J. Veness, M.G. Bellemare, A. Graves, M. Riedmiller, A.K. Fidjeland, G. Ostrovski, S. Petersen, C. Beattie, A. Sadik, I. Antonoglou, H. King, D. Kumaran, D. Wierstra, S. Legg, D. Hassabis, Human-level control through deep reinforcement learning. Nature **518**(7540), 529–533 (2015)
40. Z. Ning, P. Dong, X. Wang, L. Guo, J.J.P.C. Rodrigues, X. Kong, J. Huang, R.Y.K. Kwok, Deep reinforcement learning for intelligent internet of vehicles: an energy-efficient computational offloading scheme. IEEE Trans. Cognit. Commun. Netw. **5**(4), 1060–1072 (2019)
41. J. Chen, Z. Wei, S. Li, B. Cao, Artificial intelligence aided joint bit rate selection and radio resource allocation for adaptive video streaming over F-rans. IEEE Wireless Commun. **27**(2), 36–43 (2020)
42. J. Du, F.R. Yu, G. Lu, J. Wang, J. Jiang, X. Chu, MEC-assisted immersive VR video streaming over terahertz wireless networks: a deep reinforcement learning approach. IEEE Internet Things J. **7**, 9517–9529 (2020)
43. J. Li, X. Zhang, J. Zhang, J. Wu, Q. Sun, Y. Xie, Deep reinforcement learning-based mobility-aware robust proactive resource allocation in heterogeneous networks. IEEE Trans. Cognit. Commun. Netw. **6**(1), 408–421 (2020)
44. S. Wang, H. Liu, P.H. Gomes, B. Krishnamachari, Deep reinforcement learning for dynamic multichannel access in wireless networks. IEEE Trans. Cognit. Commun. Netw. **4**(2), 257–265 (2018)
45. J. Chen, S. Chen, Q. Wang, B. Cao, G. Feng, J. Hu, iRAF: a deep reinforcement learning approach for collaborative mobile edge computing IoT networks. IEEE Internet Things J. **6**(4), 7011–7024 (2019)
46. L. Huang, S. Bi, Y.J. Zhang, Deep reinforcement learning for online computation offloading in wireless powered mobile-edge computing networks. IEEE Trans. Mobile Comput. **19**, 2581–2593 (2019)
47. J. Schulman, F. Wolski, P. Dhariwal, A. Radford, O. Klimov, Proximal policy optimization algorithms (Online). CoRR, vol. abs/1707.06347 (2017). http://arxiv.org/abs/1707.06347
48. Y. Yu, Mobile edge computing towards 5g: vision, recent progress, and open challenges. China Commun. **13**(2), 89–99 (2016)

Chapter 5
Energy-Efficient Multi-Task Multi-Access Computation Offloading via NOMA

5.1 System Model and Problem Formulation

In this section, we firstly briefly introduce the motivation of our work in Sect. 5.1.1. We then present our considered system model of multi-access multi-task computation offloading via NOMA in Sect. 5.1.2. The detailed problem formulation of our energy efficient design of multi-access multi-task offloading via NOMA transmission is illustrated in Sect. 5.1.3.

5.1.1 Motivation

The growing developments of wireless networks and mobile Internet services have yielded a variety of computation-intensive yet delay-sensitive applications, e.g., virtual/augmented reality, autonomous driving, and real-time machine learning assisted services. However, due to the cost and hardware issues, conventional wireless devices are usually equipped with limited computation-resources, leading to a poor performance when running the computation-intensive tasks. Thanks to the recent advances in multi-access radio networks, the paradigm of multi-access mobile edge computing (MA-MEC) has provided a promising approach to address this issue [1]. With MA-MEC, a wireless device can offload part of its computation-tasks to several nearby edge-computing servers equipped with sufficient computation-resources, which effectively reduces the latency in completing the tasks. The advantage of MA-MEC and its applications have attracted lots of research interests [2–5]. In particular, the joint management of communication and computation resources plays a crucial role to the performance of computation offloading, e.g., energy-efficiency [6, 7]. Thanks to the recent advances in deep learning, many research efforts have been devoted to exploiting deep learning for computation offloading and the associated resource management [8–14].

Y. Chen et al., *Energy Efficient Computation Offloading in Mobile Edge Computing*, Wireless Networks, https://doi.org/10.1007/978-3-031-16822-2_5

Due to the need of delivering the offloaded workloads to several edge-computing servers, the success of MA-MEC necessitates an efficient multi-access wireless transmission scheme. Non-orthogonal multiple access (NOMA), which allows the wireless device to simultaneously send its offloaded workloads to several edge-computing servers over a same frequency channel, has provided a highly efficient approach to realize the parallel offloading in MA-MEC. Compared to conventional orthogonal multiple access (OMA), NOMA avoids the orthogonal separation of radio resources and exploits successive interference cancellation (SIC) to mitigate the consequent co-channel interference. As a result, NOMA can improve the efficiency in radio resource utilization and enhance the system throughput [15–17]. In particular, there have been several studies investigating the NOMA enabled edge computing [18–24].

In practice, a wireless device may execute a group of tasks simultaneously, with each task having a latency-limit to be completed [25–27]. Thus, from the perspective of improving energy-efficiency, it is a crucial question about how to properly offload the workloads of different tasks to the edge-computing servers. In addition, the optimal offloading solution depends on the channel power gains from the wireless device to the edge-computing servers. Considering the effect of time-varying channels, the huge number of different channel realizations (across the time-horizon) almost prohibit us from using conventional optimization approach to determine the optimal offloading solution for each channel realization experienced in realtime. Motivated by these challenges, in this work, we thus investigate the NOMA assisted MA-MEC for multi-task computation offloading via deep reinforcement learning.

The contributions in this chapter can be summarized as follows. *First*, the chapter investigates the NOMA assisted MA-MEC under a static channel scenario, in which a wireless device is running a group of tasks. The device can offload parts of the workloads of the tasks to a group of edge-computing servers via NOMA. To study this problem, the chapter formulates a joint optimization of the multi-task computation offloading, NOMA transmission, and computation-resource allocation, with the objective of minimizing the total energy consumption of the wireless device. Although the formulated joint optimization problem is strictly non-convex, we identify its layered structure and propose a layered algorithm to find the optimal offloading solution. *Second*, this chapter considers the dynamic channel scenario in which the channel power gains from the wireless device to the edge-computing servers are time-varying. To tackle with the difficulty due to the huge number of different channel realizations in the dynamic scenario, an online algorithm based on Deep Reinforcement Learning (DRL) is proposed to efficiently obtain the near-optimal offloading solutions for different channel realizations. In particular, by exploiting the layered feature of the joint optimization problem before, the proposed DRL algorithm implements a deep neural network (DNN) to learn the optimal NOMA-transmission duration (i.e., a scalar) according to different channel realizations from the historical experiences. Thus, the proposed online algorithm avoids solving the non-convex joint optimization for each individual channel realization and thus reduces the complexity in finding the offloading solutions under the time-varying channel scenario. Finally, numerical results are provided to validate the performance of the proposed algorithms. Part contents in this chapter have been published in [28].

5.1.2 System Model

Figure 5.1 shows the system model considered in this chapter, where one wireless device (WD) is running a group of tasks denoted by $\mathcal{K} = \{1, 2, \ldots, K\}$. For each task $k \in \mathcal{K}$, we use S_k^{tot} to denote its total workload to be completed and use $T_k^{\max}$ to denote the latency-limit. There exist a group of edge-computing servers (ECSs) denoted by $\mathcal{I} = \{1, 2, \ldots, I\}$, with ECS i co-located with the i-th base station (BS i). To facilitate the model of NOMA transmission, we assume

$$g_1 \geq g_2 \geq \ldots \geq g_I, \tag{5.1}$$

where g_i denotes the channel power gain from the WD to ECS i.

The WD can divide the total workloads of task k as $\mathbf{s}_k = [s_{k1}, s_{k2}, \ldots, s_{kI}]$, with s_{ki} denoting the amount of the computation-workloads offloaded to ECS i.

Given the WD's multi-access offloading decision $\{\mathbf{s}_k\}_{k\in\mathcal{K}}$, the total amount of the offloaded workloads to ECS i can be expressed as $\sum_{k\in\mathcal{K}} s_{ki}$. We use t to denote the duration of NOMA transmission in which the WD delivers $\{\mathbf{s}_k\}_{k\in\mathcal{K}}$ to the respective

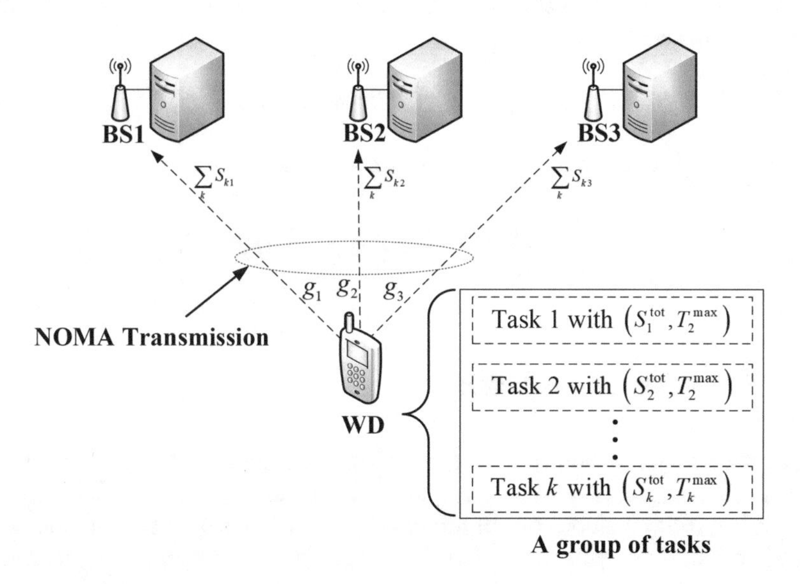

Fig. 5.1 WD's multi-access computation-offloading via NOMA: A WD offloads its computational workloads from a group of tasks to three edge-computing servers simultaneously via non-orthogonal multiple access

ECSs. Thus, for sending $\{\mathbf{s}_k\}_{k\in\mathcal{K}}$ to the ECSs in t, we can express the WD's required minimum transmit-power as follows:

$$P^{\text{NOMA}}(\{\mathbf{s}_k\}_{k\in\mathcal{K}}, t) = Wn_0 \sum_{i=1}^{I} (\frac{1}{g_i} - \frac{1}{g_{i-1}}) 2^{\frac{1}{t}\frac{1}{W}\sum_{m=i}^{I}(\sum_{k=1}^{K} s_{km})} - \frac{Wn_0}{g_I}, \quad (5.2)$$

where W denotes the channel bandwidth, and n_0 denotes the power spectral density of the background noise.

To model the latency of the WD in completing all its tasks, we first introduce $\boldsymbol{\mu}_{\text{L}} = [\mu_{1\text{L}}, \mu_{2\text{L}}, \ldots, \mu_{K\text{L}}]$ to denote the WD's local computation-rates for processing the remaining workloads of the tasks, with $\mu_{k\text{L}}$ denoting the local computation-rate for task k. Furthermore, we introduce $\boldsymbol{\gamma}_i = [\gamma_{1i}, \gamma_{2i}, \ldots, \gamma_{Ki}]$ to denote the ECS i's computation-rates for processing the offloaded workloads, with γ_{ki} denoting the allocated computation-rate for task k. We call $\boldsymbol{\mu}_{\text{L}}$ and $\{\boldsymbol{\gamma}_i\}_{i\in\mathcal{I}}$ as the computation-resource allocations.

Based on the above modeling, for each task k, we can express its overall-latency as

$$d_k = \max\left\{\frac{S_k^{\text{tot}} - \sum_{i\in\mathcal{I}} s_{ki}}{\mu_{k\text{L}}}, \max_{i\in\mathcal{I}}\{t + \frac{s_{ki}}{\gamma_{ki}}\}\right\}, \forall k \in \mathcal{K}. \quad (5.3)$$

As the WD is energy-capacity limited, we aim at minimizing the total energy consumption of the WD to complete all its tasks. Specifically, the total energy consumption of the WD includes the energy consumption E_{NOMA} for the offloading transmission, i.e.,

$$E_{\text{NOMA}} = tP^{\text{NOMA}}(\{\mathbf{s}_k\}_{k\in\mathcal{K}}, t), \quad (5.4)$$

and the energy consumption E_{LC} for its local computing, i.e.,

$$E_{\text{LC}} = \sum_{k\in\mathcal{K}} \frac{S_k^{\text{tot}} - \sum_{i\in\mathcal{I}} s_{ki}}{\mu_{k\text{L}}} \rho_{\text{L}} \mu_{k\text{L}}^3 = \rho_{\text{L}} \sum_{k\in\mathcal{K}} (S_k^{\text{tot}} - \sum_{i\in\mathcal{I}} s_{k,i}) \mu_{k\text{L}}^2. \quad (5.5)$$

5.1.3 Problem Formulation

We formulate a joint optimization of the NOMA-transmission duration t, the multi-access offloading decision $\{\mathbf{s}_k\}_{k\in\mathcal{K}}$, and the computation-resource allocations $\boldsymbol{\mu}_{\text{L}}$ and $\{\boldsymbol{\gamma}_i\}_{i\in\mathcal{I}}$, with the objective of minimizing the total energy consumption of the WD. The details are as follows ("TEM" standard for "Total Energy Minimization"):

$$\text{(TEM):} \quad \min E_{\text{NOMA}} + E_{\text{LC}}$$

$$\text{subject to:} \quad P^{\text{NOMA}}(\{\mathbf{s}_k\}_{k\in\mathcal{K}}, t) \le P^{\max}, \quad (5.6)$$

$$d_k \leq T_k^{\max}, \forall k \in \mathcal{K}, \tag{5.7}$$

$$\sum_{k \in \mathcal{K}} \mu_{k\mathrm{L}} \leq \mu_{\mathrm{L}}^{\max}, \tag{5.8}$$

$$\sum_{k \in \mathcal{K}} \gamma_{ki} \leq \gamma_i^{\max}, \forall i \in \mathcal{I}, \tag{5.9}$$

$$\sum_{i \in \mathcal{I}} s_{ki} \leq S_k^{\mathrm{tot}}, \forall k \in \mathcal{K}, \tag{5.10}$$

$$\text{variables:} \quad t \geq 0, \{\mathbf{s}_k\}_{k \in \mathcal{K}}, \boldsymbol{\mu}_{\mathrm{L}}, \text{ and } \{\boldsymbol{\gamma}_i\}_{i \in \mathcal{I}}.$$

In Problem (TEM), constraint (5.6) ensures that the transmit-power of the WD cannot exceed the power capacity $P^{\max}$ of its radio interface. Constraint (5.7) ensures that the overall-latency for completing task k cannot exceed the latency-limit $T_k^{\max}$. Constraint (5.8) ensures that the sum of the WD's allocated computation-rates for all tasks cannot exceed its local computation-rate capacity denoted by $\mu_{\mathrm{L}}^{\max}$, and constraint (5.9) means that for each ECS i, the sum of its allocated computation-rates for different tasks cannot exceed ECS i's computation-rate capacity denoted by $\gamma_i^{\max}$.

However, Problem (TEM) is a strictly non-convex optimization problem. Thus, there exists no general algorithm that can efficiently solve Problem (TEM). To address this difficulty, we exploit the layered structure of Problem (TEM) and propose a corresponding layered algorithm for solving it. The details are presented in the next section.

5.2 LEEMMO: Layered Energy-Efficient Multi-Task Multi-Access Algorithm

In this section, we present our algorithmic design for solving Problem (TEM). We firstly perform a series of equivalent transformations of Problem (TEM), based on which we identify its layered structure in Sect. 5.2.1. Specifically, in our proposed layered structure, we propose a subproblem that aims at optimizing the computation-offloading decisions. With the output of the subproblem, we propose the top-problem aims at optimizing the NOMA transmission duration. Based on the layered structure, in Sect. 5.2.2, we firstly propose a subroutine for solving the subproblem. After that, in Sect. 5.2.3, we propose a top-layer algorithm, which exploits the previously designed subroutine, for solving the top-problem. Finally, in Sect. 5.2.4, we consider the time-varying channel states between the WD and ECSs and propose a DRL based algorithm for solving Problem (TEM) under the dynamic conditions.

5.2.1 Layered Decomposition of Joint Optimization Problem

With (5.3), we can transform (5.7) into:

$$S_k^{\text{tot}} - \mu_{k\text{L}} T_k^{\max} \leq \sum_{i \in \mathcal{I}} s_{ki}, \forall k \in \mathcal{K}, \tag{5.11}$$

and

$$s_{ki} \leq \gamma_{ki}(T_k^{\max} - t), \forall k \in \mathcal{K}, \forall i \in \mathcal{I}. \tag{5.12}$$

With (5.11), we can equivalently transform constraint (5.8) into:

$$\sum_{k \in \mathcal{K}} \frac{S_k^{\text{tot}} - \sum_{i \in \mathcal{I}} s_{ki}}{T_k^{\max}} \leq \mu_{\text{L}}^{\max}. \tag{5.13}$$

Meanwhile, with (5.12), we can transform constraint (5.9) into:

$$\sum_{k \in \mathcal{K}} \frac{s_{ki}}{T_k^{\max} - t} \leq \gamma_i^{\max}, \forall i \in \mathcal{I}. \tag{5.14}$$

Further with (5.11), the WD's total energy consumption for its local computing can be expressed as:

$$E_{\text{LC}} = \rho_{\text{L}} \sum_{k \in \mathcal{K}} (S_k^{\text{tot}} - \sum_{i \in \mathcal{I}} s_{ki})^3 \frac{1}{(T_k^{\max})^2}.$$

Based on the above operations, we obtain the following Equivalent form of Problem (TEM):

$$\text{(TEM-E):} \quad \min \sum_{k \in \mathcal{K}} \rho_{\text{L}} (S_k^{\text{tot}} - \sum_{i \in \mathcal{I}} s_{ki})^3 \frac{1}{(T_k^{\max})^2} + t P^{\text{NOMA}}(\{s_{ki}\}_{k \in \mathcal{K}, i \in \mathcal{I}}, t)$$

$$\text{subject to:} \quad \text{constraints (5.6), (5.10), (5.13), and (5.14),}$$

$$\text{variables:} \quad t \geq 0, \text{and} \{\mathbf{s}_k \geq 0\}_{k \in \mathcal{K}}.$$

Problem (TEM-E) is still a strictly non-convex optimization problem. To solve it efficiently, we propose the following decomposition approach.

(Subproblem under given t): Suppose that t is given in advance. We solve the following subproblem:

$$\text{(TEM-E-Sub):} \quad E_{\text{tot,sub}}(t) = \min \sum_{k \in \mathcal{K}} \rho_{\text{L}} \frac{(S_k^{\text{tot}} - \sum_{i \in \mathcal{I}} s_{ki})^3}{(T_k^{\max})^2} + t P^{\text{NOMA}}(\{\mathbf{s}_k\}_{k \in \mathcal{K}}, t)$$

$$\text{subject to:} \quad P^{\text{NOMA}}(\{\mathbf{s}_k\}_{k\in\mathcal{K}}, t) \leq P^{\max}, \tag{5.15}$$

$$\sum_{k\in\mathcal{K}} \frac{s_{ki}}{T_k^{\max} - t} \leq \gamma_i^{\max}, \forall i \in \mathcal{I}, \tag{5.16}$$

$$\text{constraints: (5.10) and (5.13),}$$

$$\text{variables:} \quad \{\mathbf{s}_k \geq 0\}_{\forall k\in\mathcal{K}}.$$

Notice that in Problem (TEM-E-Sub), the value of t is already given in (5.15) and (5.16).

(Top-problem to optimize t***):*** After solving Problem (TEM-E-Sub) and obtaining the value of $E_{\text{tot,sub}}(t)$ (for each given value of t), we then continue to optimize the value of t, which corresponds to the following optimization problem:

$$\text{(TEM-E-Top):} \quad \min E_{\text{tot,sub}}(t)$$

$$\text{variable:} \quad 0 \leq t \leq \min_{k\in\mathcal{K}}\{T_k^{\max}\}.$$

5.2.2 *Proposed Subroutine for Solving Problem (TEM-E-Sub)*

In this subsection, we propose a subroutine to solve Problem (TEM-E-Sub) for each given value of t. Specifically, given the value of t, we firstly identify the following important feature.

Proposition 5.1 *Problem (TEM-E-Sub) is a strictly convex optimization problem with respect to* $\{\mathbf{s}_k\}_{\forall k\in\mathcal{K}}$.

Proof Although $P^{\text{NOMA}}(\{\mathbf{s}_k\}_{k\in\mathcal{K}}, t)$ given in Eq. (5.2) is not jointly convex with respect to $\{\mathbf{s}_k\}_{\forall k\in\mathcal{K}}$ and t, an important feature is that $P^{\text{NOMA}}(\{\mathbf{s}_k\}_{k\in\mathcal{K}}, t)$ is strictly convex with respect to $\{\mathbf{s}_k\}_{k\in\mathcal{K}}$ when the value of t is fixed in advance. Moreover, it can also be identified that the first item in the objective function of Problem (TEM-E-Sub), i.e., $\sum_{k\in\mathcal{K}} \rho_{\text{L}}(S_k^{\text{tot}} - \sum_{i\in\mathcal{I}} s_{ki})^3 \frac{1}{(T_k^{\max})^2}$, is strictly convex with respect to $\{\mathbf{s}_k\}_{k\in\mathcal{K}}$. As a result, given the value of t, the objective function of Problem (TEM-E-Sub) is strictly convex in $\{\mathbf{s}_k\}_{\forall k\in\mathcal{K}}$. Meanwhile, with the given value of t, constraint (5.15) leads to a convex feasible set for $\{\mathbf{s}_k\}_{k\in\mathcal{K}}$, and constraints (5.10), (5.13), and (5.16) are all affine in $\{\mathbf{s}_k\}_{k\in\mathcal{K}}$. As a result, it can be concluded that Problem (TEM-E-Sub) is a strictly convex optimization problem with respect to $\{\mathbf{s}_k\}_{\forall k\in\mathcal{K}}$.

Exploiting Proposition 5.1, we propose an algorithm to solve Problem (TEM-E-Sub) as follows.

We introduce the dual variables $\mathbf{z} = [z_1, z_2, \ldots, z_I]$ for relaxing constraint (5.16), which leads to the following Lagrangian function:

$$L(\{\mathbf{s}_k\}_{\forall k\in\mathcal{K}}, \mathbf{z}) = \sum_{k\in\mathcal{K}} \rho_{\mathrm{L}}(S_k^{\mathrm{tot}} - \sum_{i\in\mathcal{I}} s_{ki})^3 \frac{1}{(T_k^{\max})^2} + t P^{\mathrm{NOMA}}(\{\mathbf{s}_k\}_{k\in\mathcal{K}}, t)$$
$$+ \sum_{i\in\mathcal{I}} z_i \sum_{k\in\mathcal{K}} \frac{s_{ki}}{T_k^{\max} - t}. \tag{5.17}$$

With (5.17) and $\mathbf{z}$, the WD optimizes its offloaded computation-workloads $\{s_k\}_{k\in\mathcal{K}}$ for minimizing its Lagrangian function, which corresponds to the following *primal optimization* problem:

$$\begin{aligned} \text{(WD-P):} \quad & \min L(\{\mathbf{s}_k\}_{\forall k\in\mathcal{K}}, \mathbf{z}) \\ \text{subject to:} \quad & \text{constraints: (5.15), (5.10), and (5.13),} \\ \text{variables:} \quad & \{\mathbf{s}_k \geq 0\}_{\forall k\in\mathcal{K}}. \end{aligned}$$

Two features of Problem (WD-P) are as follows. Given $\mathbf{z}$, the WD can individually solve Problem (WD-P), which is a convex optimization problem. Thus, the WD can solve Problem (WD-P) by using the conventional solver for convex optimizations (e.g., CVX).

Let us use $\{\mathbf{s}_k^{\mathrm{o}}\}_{\forall k\in\mathcal{K}}$ to denote the optimal solution of Problem (WD-P). With $\{\mathbf{s}_k^{\mathrm{o}}\}_{\forall k\in\mathcal{K}}$, the group of the ECSs aim at solving the following *dual optimization* problem:

$$\begin{aligned} \text{(ECS-P):} \quad & \max L(\{\mathbf{s}_k^{\mathrm{o}}\}_{\forall k\in\mathcal{K}}, \mathbf{z}) \\ \text{subject to:} \quad & \{z_i \geq 0\}_{i\in\mathcal{I}}. \end{aligned}$$

In particular, Problem (ECS-P) can be solved by in a distributed manner as follows:

$$z_i = \left(z_i + \varrho \Big(\sum_{k\in\mathcal{K}} \frac{s_{ki}^{\mathrm{o}}}{T_k^{\max} - t} - \gamma_i^{\max} \Big) \right)^+, \forall i \in \mathcal{I}, \tag{5.18}$$

where ϱ is the step-size for dual updating, and function $(x)^+ = \max\{x, 0\}$. Specifically, given the WD's offloaded workloads from all tasks to ECS i, ECS i can individually update its dual variable z_i according to (5.18), without requiring any information from other ECSs.

Exploiting (5.18), we propose an algorithm for solving Problem (TEM-E-Sub). The details of the subroutine are shown in Algorithm 1. Specifically, in Step 4, the WD updates its offloaded workloads $\{\mathbf{s}_k\}_{k\in\mathcal{K}}$ by solving Problem (ECS-P) with the given $\mathbf{z}$. Then, each ECS i updates its z_i in Step 9 based on the offloaded workloads from the WD. The iteration process continues until there is no change in the WD's offloaded workloads in Step 6. An important advantage of our subroutine for solving Problem (TEM-E-Sub) is that it does not require the WD and the ECSs to

Algorithm 1 Subroutine for solving Problem (TEM-E-Sub) and obtaining the value of $E_{\text{tot,sub}}(t)$

1: **Input:** The value of t.
2: Each ECS i initializes its $z_i \geq 0$ and sends z_i to the WD.
3: **while** (1) **do**
4: After receiving $\{z_i\}_{i \in \mathcal{I}}$, the WD solves Problem (WD-P) (e.g., by using CVX) and obtains the corresponding computation-workloads $\{\mathbf{s}_k^{\text{o}}\}_{\forall k \in \mathcal{K}}$ to be offloaded.
5: **if** There is no change in $\{\mathbf{s}_k^{\text{o}}\}_{\forall k \in \mathcal{K}}$ compared to the previous round of iteration **then**
6: Convergence is reached and go to Step 11.
7: **end if**
8: To each ECS i, the WD sends the profile of $\{s_{ki}^{\text{o}}\}_{k \in \mathcal{K}}$.
9: Each ECS i, after receiving $\{s_{ki}^{\text{o}}\}_{k \in \mathcal{K}}$ from the WD, updates its z_i according to (5.18).
10: **end while**
11: **Output:** The WD uses $\{s_{ki}^{\text{o}}\}_{k \in \mathcal{K}}$ as the optimal solution for Problem (TEM-E-Sub) and computes the value of $E_{\text{tot,sub}}(t)$.

reveal their respective local information to each other, which thus helps preserve the privacy of the WD and the ECSs. Specifically, in our proposed subroutine, the WD only needs to determine its offloaded computation-workloads to the ECSs according to the dual variables from the ECSs, and there is no need for the WD to disclose its local information (e.g., $\{S_k^{\text{tot}}, T_k^{\max}\}_{k \in \mathcal{K}}$ and $\mu_{\text{L}}^{\max}$) to the ECSs. Similarly, each ECS i only needs to update its dual variable z_i according to the offloaded computation-workloads from the WD, and there is no need for each ECS i to disclose its local information (e.g., $\gamma_i^{\max}$) to the WD.

5.2.3 A Layered Algorithm for Solving Problem (TEM-E-Top)

Until now, we can use the proposed subroutine to solve Problem (TEM-E-Sub) and obtain the value of $E_{\text{tot,sub}}(t)$ for each given value of t. We next solve Problem (TEM-E-Top). Although we cannot express $E_{\text{tot,sub}}(t)$ analytically, Problem (TEM-E-Top) can be solved efficiently, since it is a single-variable optimization problem with t falling within a fixed interval. We thus propose Top-Algorithm, which executes a line-search with a small stepsize Δ, to numerically solve Problem (TEM-E-Top). The details of our Top-Algorithm are presented in Algorithm 2.

In Top-Algorithm, in each round of the iterations, given the currently evaluated t, the WD and ECSs invoke the subroutine to obtain the value of $E_{\text{tot,sub}}(t)$. If $E_{\text{tot,sub}}(t)$ is smaller than the current best value (CBV), the WD updates the current best solution (CBS) in Step 6. After completing Top-Algorithm, the WD can obtain its optimal transmission duration t^* and the corresponding multi-access offloading decisions $\{\mathbf{s}_k^*\}_{k \in \mathcal{K}}$.

Finally, with t^* and $\{\mathbf{s}_k^*\}_{k \in \mathcal{K}}$, we can derive the WD's optimal local computation-rate allocations for all tasks as:

Algorithm 2 Top-algorithm for solving problem (TEM-E)

1: The WD sets a small step-size Δ and initializes $t = \Delta$. The WD sets the current best value (CBV) as a vary large number, and sets the current best solution (CBS) as an empty set.
2: **while** $t \leq \min_{k\in\mathcal{K}}\{T_k^{\max}\}$ **do**
3: The WD and ECSs together invoke the subroutine to solve Problem (TEM-E-Sub), and the WD obtains the value of $E_{\text{tot,sub}}(t)$ as well as the corresponding $\{s_{ki}^{\text{o}}\}_{k\in\mathcal{K},i\in\mathcal{I}}$.
4: **if** $E_{\text{tot,sub}}(t) < \text{CBV}$ **then**
5: The WD updates $\text{CBV} = E_{\text{tot,sub}}(t)$.
6: The WD updates $\text{CBS} = \{t, \{s_{ki}^{\text{o}}\}_{k\in\mathcal{K},i\in\mathcal{I}}\}$.
7: **end if**
8: The WD updates $t = t + \Delta$.
9: **end while**
10: **Output:** Based on CBV, set $t^* = t$ and $\{s_{ki}^* = s_{ki}^{\text{o}}\}_{k\in\mathcal{K},i\in\mathcal{I}}$ as the optimal solution for Problem (TEM-E).

$$\mu_{k\text{L}}^* = \frac{S_k^{\text{tot}} - \sum_{i\in\mathcal{I}} s_{ki}^*}{T_k^{\max}}, \forall k \in \mathcal{K}, \tag{5.19}$$

and each the ECS's computation-rate allocations for all tasks as:

$$\gamma_{ki}^* = \frac{s_{ki}^*}{T_k^{\max} - t^*}, \forall k \in \mathcal{K}, i \in \mathcal{I}. \tag{5.20}$$

Thus, we complete solving Problem (TEM).

As a summary of the above layered algorithmic design, we present Fig. 5.2 to demonstrate the connection between Top-Algorithm and its subroutine. As shown in Fig. 5.2, given the enumerated value of variable t (i.e., the NOMA transmission), our previously proposed subroutine is invoked to solve Problem (TEM-Sub-E) and provide the corresponding value of $E_{\text{tot,sub}}(t)$. Then, with the output value of $E_{\text{tot,sub}}(t)$, our Top-Algorithm executes a line-search on the value of t to solve Problem (TEM-Top), with the objective of finding the best value of t that can achieve the minimum of $E_{\text{tot,sub}}(t)$. This completely solves our original Problem (TEM).

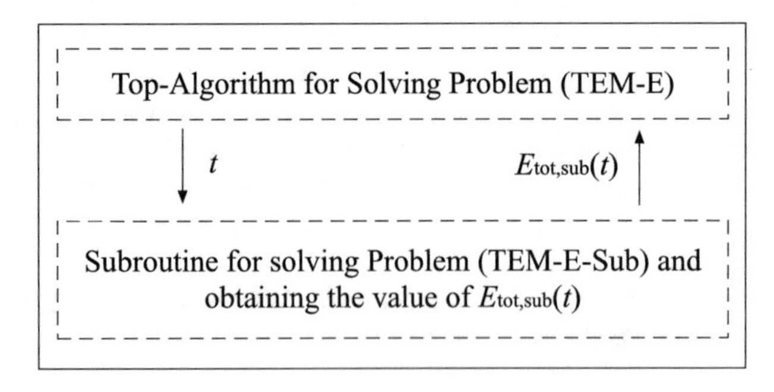

Fig. 5.2 An illustration of our layered algorithmic design with Top-Algorithm to solve problem (TEM-Top) and its subroutine to solve problem (TEM-Sub-E) for each enumerated value of t

5.2.4 DRL-Based Online Algorithm

Our proposed Top-Algorithm and its subroutine mainly aim at solving Problem (TEM) under a relatively static scenario, in which the channel states between the WD and the ECSs keep unchanged. In practice, the channel states between the WD and the ECSs may vary over time, leading to a huge number of possible channel realizations (i.e., different vectors $\mathbf{g} = [g_1, g_2, \ldots, g_I]$) at different time slots. As a result, it is challenging for us to adopt Top-Algorithm (and its subroutine) in a real-time manner to solve Problem (TEM) for each experienced channel realization. To address this difficulty, we propose an online algorithm, which is based on DRL, to quickly learn the near-optimal solution for Problem (TEM) at different time slots.

In particular, the DRL based online algorithm has been envisioned as a promising approach for dynamic programming problems with complicated system space and action space. Taking into account the dynamic network conditions (e.g., time-varying channel conditions and user demands), dynamic programming (DP) has been widely used for optimizing the long-term decision process. Conventional reinforcement learning (RL) based algorithms provide effective approaches for solving different types of DP problems, including both the finite-horizon DP problems as well as the infinite-horizon DP problems. Nevertheless, with the growing dimension (e.g., the system space and action space) of the DP problems, conventional RL may suffer from a low convergence rate, since it invokes an iterative approach (e.g., the value iteration and policy iteration) to compute the so-called Q-values for all possible state-action pairs. Thus, the larger sizes of the system space and action space usually result in a larger computational complexity (or a longer latency) for reaching the convergence of the iterative approach. Facing the fast varying of the network conditions (e.g., the variations in radio channels), it might become prohibitive to use RL for solving large-scale DP problems directly. Moreover, for the ease of the using the iterative computation to compute the Q-values of all state-action pairs, conventional RL based algorithms usually require a discrete action space. Such a feature, however, does not fit lots of DP problems with continuous action space. To address the aforementioned issues, leveraging the recent advanced deep learning via DNN into traditional RL provides a promising solution, which yields the solution methodology of DRL. The key essence of DRL is to use DNN to directly approximate Q-values of the state-action pairs, which thus can avoid the traditional approach of iterative computation. The advantage of using DNN lies in that it can provide the approximated Q-values of the state-action pairs in a very efficient manner (based on the historical experiences), and moreover, it can accommodate the action with a continuous space since no Q-table is involved. Nevertheless, there also exist some disadvantages when using DRL-based algorithm for addressing the DP problems. For instance, to accurately approximate the Q-values of the state-action pairs, a certain amount of samples (namely, the historical rewards when selecting the actions under different system states) are required for training the DNN, which thus invokes a certain amount of overheads. In addition,

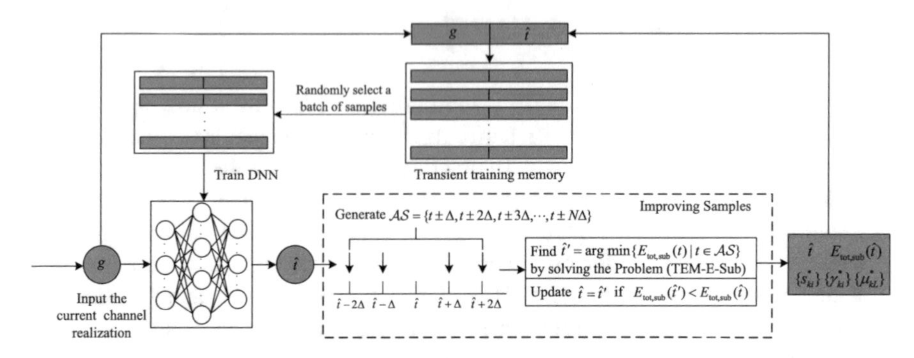

Fig. 5.3 Framework of our proposed DRL-based online algorithm

due to the nature of approximation via DNN, using DRL based algorithm cannot be guaranteed to achieve the globally optimal solution to the DP problems.

Exploiting the layered structure of Problem (TEM) discussed in Sect. 5.2.1, our DRL algorithm implements a DNN to learn the optimal NOMA-transmission duration t according to different channel realizations $\mathbf{g}$ from the historical experience. Thus, we treat $\mathbf{g}$ as the system state and treat t as the action. Meanwhile, we use the WD's total energy consumption $E_{\text{tot,sub}}(t)$ as the reward. Our objective is to train the DNN to learn the optimal NOMA-transmission duration t^* for minimizing the WD's total energy consumption, which corresponds to finding the following policy π:

$$t^* = \pi_{\boldsymbol{\theta}}(\mathbf{g}). \tag{5.21}$$

Here, $\boldsymbol{\theta}$ denotes the learning-parameters used in the DNN module. Figure 5.3 shows a framework of our DRL-based online algorithm. With the learned value of $t^* = \pi_{\boldsymbol{\theta}}(\mathbf{g})$, we can further obtain the corresponding multi-access offloading decision and computation-resource allocations by solving Problem (TEM-E-Sub). Such an approach (i.e., learning the scalar t^*) can effectively improve the accuracy and efficiency of the DNN module, in comparison with directly learning the whole set of the solutions of Problem (TEM).

As in Fig. 5.3, our DRL-based algorithm includes two parts, i.e., improving the sample and updating policy π.

- *(Improving the samples)* A key feature of our DRL-based algorithm is that we propose an additional-sampling scheme to improve its accuracy. The details are as follows. In each time slot, given the current channel realization $\mathbf{g}$, the DNN yields $\hat{t} = \pi_{\boldsymbol{\theta}}(\mathbf{g})$ with the policy $\pi_{\boldsymbol{\theta}}$. However, $\hat{t}$ may not be optimal to Problem (TEM-E-Top). To improve $\hat{t}$, we propose the additional-sampling (AS) scheme, namely, using $\hat{t}$ as the central, we uniformly take a group of additional samples as $\mathcal{AS} = \{\hat{t} - N\Delta, \ldots, \hat{t} - 2\Delta, \hat{t} - \Delta, \hat{t} + \Delta, \hat{t} + 2\Delta, \ldots, \hat{t} + N\Delta\}$, where Δ is the sampling space, and $2N$ is the number of the additional samples. Within $\mathcal{AS}$, we find $\hat{t}' = \arg\min\{E_{\text{tot,sub}}(t) | t \in \mathcal{AS}\}$. We then update $\hat{t} = \hat{t}'$ if $E_{\text{tot,sub}}(\hat{t}') <$

$E_{\text{tot,sub}}(\hat{t})$. Finally, we obtain $(\mathbf{g}, \hat{t})$ as a new sample of the channel realization and the transmission-duration to be stored in the training memory.

- *(Updating the policy)* After collecting a certain number of new samples, we randomly select a batch of the samples from the training memory, and use them to train the DNN (i.e., updating the learning-parameters $\boldsymbol{\theta}$), which consequently yields an updated policy. In particular, let $\{(\mathbf{g}_\tau, t_\tau)\}_{\tau \in \mathcal{T}}$ denote the randomly selected batch of the samples from the training memory, where $\mathcal{T}$ is a set of indexes of the selected samples. The learning-parameters $\boldsymbol{\theta}$ inside of the DNN module can be adjusted by Adam algorithm, i.e., minimizing the following mean-square error function as

$$\mathcal{L}(\boldsymbol{\theta}) = \frac{1}{|\mathcal{T}|} \sum_{\tau \in \mathcal{T}} \left(\pi_{\boldsymbol{\theta}}(\mathbf{g}_\tau) - \hat{t}_\tau\right)^2, \tag{5.22}$$

where $|\mathcal{T}|$ is the number of the selected samples for training.

By iteratively executing the above two parts, the DNN trains policy $\pi_{\boldsymbol{\theta}}$ which can eventually provide the near-optimal values of t to Problem (TEM). The details of our proposed DRL-based online algorithm are shown in Algorithm 3.

Algorithm 3 DLR-based online algorithm to yield the near-optimal solution for Problem (TEM)

1: **Input:** Generate a group of channel realizations $\{\mathbf{g}_\tau\}_{\tau \in \mathcal{T}}$. For each channel realization $\mathbf{g}_\tau$, use Top-Algorithm to obtain the corresponding $\hat{t}_\tau$. Store the group of the samples $\{(\mathbf{g}_\tau, \hat{t}_\tau)\}_{\tau \in \mathcal{T}}$ into the training memory.
2: Initialize the time index $l = 1$, and the learning-parameters $\boldsymbol{\theta}$ and the corresponding mapping policy $\pi_{\boldsymbol{\theta}}$.
3: **while** Convergence condition is not reached **do**
4:     **if** l mod $N_{\text{rt}} = 0$ **then**
5:         Randomly select a batch of samples $\{(\mathbf{g}_\tau, \hat{t}_\tau)\}_{\tau \in \mathcal{T}}$ from the training memory.
6:         Update $\boldsymbol{\theta}$ in DNN by using Adam Algorithm with the selected samples, and obtain the updated policy π.
7:     **end if**
8:     Generate $\hat{t}_l = \pi_{\boldsymbol{\theta}}(\mathbf{g}_l)$ with the current $\mathbf{g}_l$,
9:     Generate the group of additional samples $\mathcal{AS}$, and find $\hat{t}' = \arg\min\{E_{\text{tot,sub}}(t) | t \in \mathcal{AS}\}$.
10:    **if** $E_{\text{tot,sub}}(\hat{t}') < E_{\text{tot,sub}}(\hat{t}_l)$ **then**
11:         Update $\hat{t}_l = \hat{t}'$.
12:    **end if**
13:    Store the sample $(\mathbf{g}_l, \hat{t}_l)$ into the training memory.
14:    Given the sample of $(\mathbf{g}_l, \hat{t}_l)$, use subroutine to solve Problem (TEM-E-Sub) and obtain $\{\mathbf{s}_k^*\}_{k \in \mathcal{K}}$. Furthermore, use (5.19) and (5.20) to compute $\boldsymbol{\mu}_{\text{L}}^*$ and $\{\boldsymbol{\gamma}_i^*\}_{i \in \mathcal{I}}$. Finally, for channel realization $\mathbf{g}_l$, output $\left(\hat{t}_l, \{\mathbf{s}_k^*\}_{k \in \mathcal{K}}, \boldsymbol{\mu}_{\text{L}}^*, \{\boldsymbol{\gamma}_i^*\}_{i \in \mathcal{I}}\right)$.
15:    Update $l = l + 1$.
16: **end while**

5.3 Performance Evaluation

In this section, we present the numerical results to validate the performance of our proposed algorithms. Specifically, we demonstrate the performance of our proposed algorithms under different parameter-settings in Sect. 5.3.1. Then, in Sect. 5.3.2, we demonstrate the performance advantage of our proposed schemes in comparison with the conventional FDMA based offloading schemes.

5.3.1 Impacts of Parameters

In this subsection, we firstly demonstrate the performance of our algorithms under different parameter-settings. Specifically, Figs. 5.4 and 5.5 demonstrate the performance of our proposed Top-Algorithm and its subroutine for solving Problem (TEM) under a static channel scenario. Then, Figs. 5.6 and 5.7 demonstrate the performance of our proposed DRL-based algorithm for the time-varying channel scenario.

In Figs. 5.4 and 5.5, we consider a scenario of three ECSs as follows. The three ECSs are uniformly located on the circle with the center at $(0, 0)$ and the radius

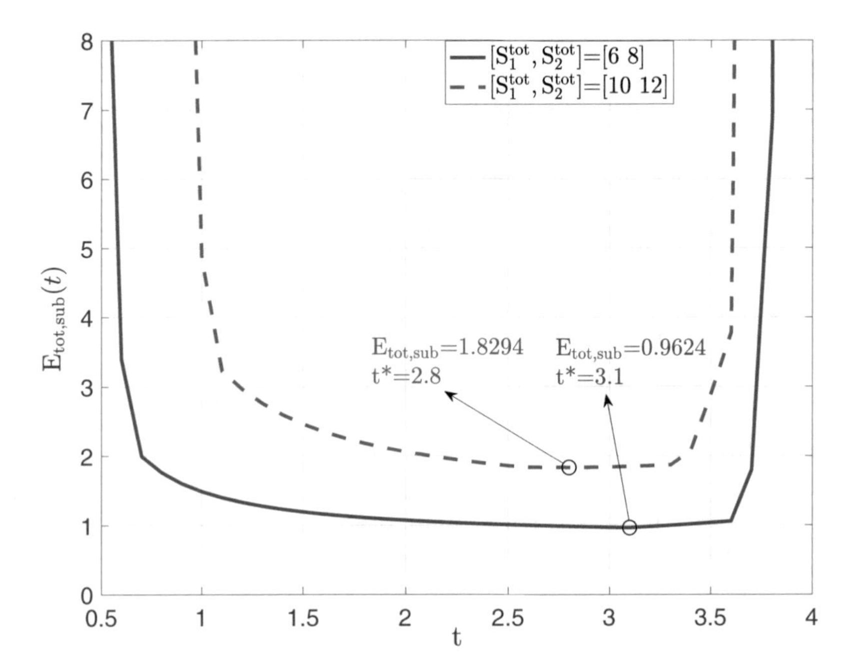

Fig. 5.4 Illustration of executing the line-search over the transmission duration t in Top-Algorithm

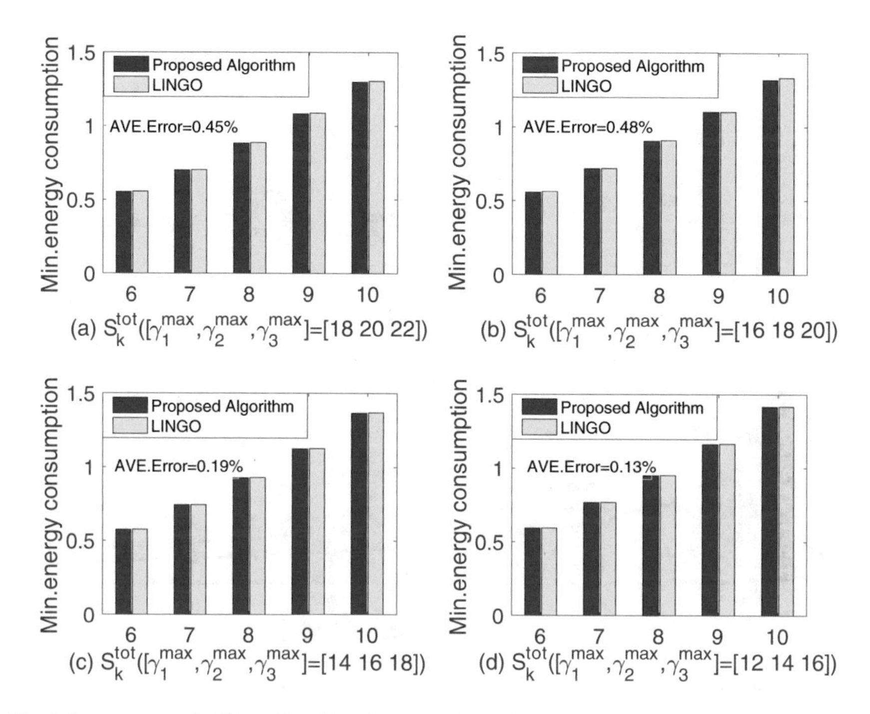

Fig. 5.5 Accuracy of of Top-Algorithm in comparison with LINGO

of 500 m. Meanwhile, the WD is randomly located within a circular plane with the center at (0, 0) and radius of 100 m, and the consequent channel power gains from the WD to the ECSs are generated according to the path-loss model with the fading effect. Based on the above settings, the randomly channel power gains, which are used in this subsection, are $\{g_i\}_{i\in\mathcal{I}} = \{2.1062, 1.3093, 0.1502\}\times 10^{-7}$. For the WD, we set its $P^{\max} = 3$W, $\mu_L^{\max} = 15$Gbits/s, and $W = 8$MHz. In addition, we set that the WD has two tasks with $[S_1^{\text{tot}}, S_2^{\text{tot}}] = [6, 8]$Mbits and $[T_1^{\max}, T_2^{\max}] = [2, 3]$ms. For the three ECSs, we set $[\gamma_1^{\max}, \gamma_2^{\max}, \gamma_3^{\max}] = [18, 19, 20]$Gbits/s.

Figure 5.4 shows the line-search over the transmission duration t in our Top-Algorithm. Specifically, we enumerate the value of t from a very small number to a large one. For each enumerated value of t, we invoke the subroutine to solve Problem (TEM-E-Sub) and obtain the corresponding value of $E_{\text{tot,sub}}(t)$. As shown in Fig. 5.4, we test two cases, namely, the case of $[S_1^{\text{tot}}, S_2^{\text{tot}}] = (6, 8)$Mbits and the case of $[S_1^{\text{tot}}, S_2^{\text{tot}}] = (10, 12)$Mbits. It can be observed from Fig. 5.4 that neither a too small transmission duration t nor a too large one will be beneficial to minimize the WD's total energy consumption. This result is consistent with the intuition. Specifically, a too small value of the transmission duration t consumes a very large transmit-power for the ED, which thus increases its energy consumption for NOMA transmission. However, a large value of the transmission duration t reduces the ED's offloaded computation workloads. As a result, the ED has to consume a large

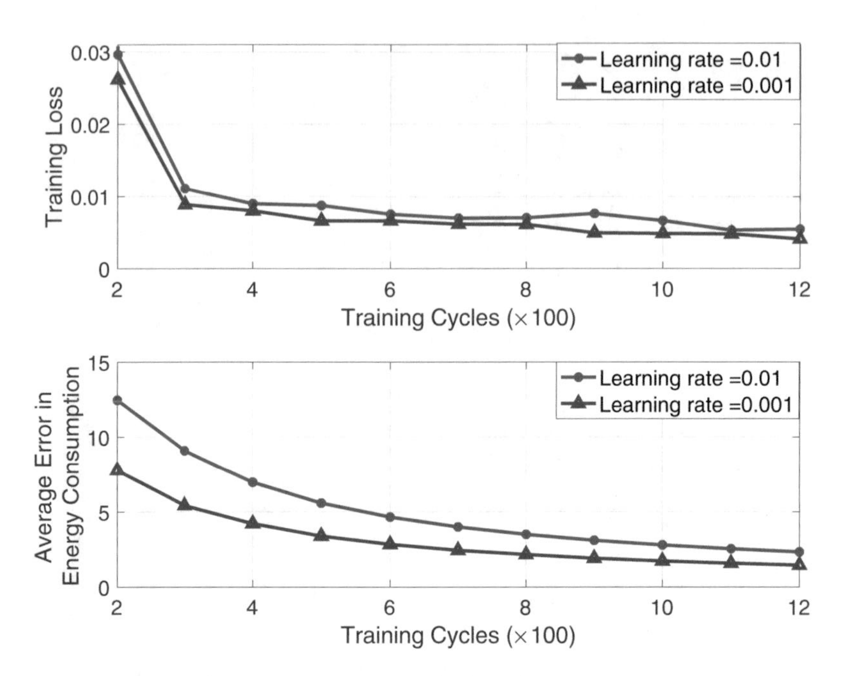

Fig. 5.6 Convergence of our proposed DRL-based online algorithm

energy consumption for processing the remaining computation workloads locally. In particular, both tested cases in Fig. 5.4 demonstrate the importance in finding the optimal t (i.e., the optimal solution of Problem (TEM-Top)) that can strike a good balance between the ED's energy consumption for its offloading transmission via NOMA and the energy consumption for the ED to process its local computation workloads.

Figure 5.5 validates the accuracy of our Top-Algorithm in solving Problem (TEM). For the purpose of validation, we use LINGO, which is a commercial optimization package, to solve Problem (TEM) directly. In particular, we use the global-solver provided by LINGO, which leverages the conventional branch and bound scheme for addressing the non-convexity of the optimization problems. Nevertheless, the disadvantage of using LINGO is that it may consume a very long computation-time. We test four cases of $[\gamma_1^{\max}, \gamma_2^{\max}, \gamma_3^{\max}]$ as shown in the four subplots in Fig. 5.5. For each tested case, we further vary S_k^{tot} from 6 Mbits to 10 Mbits. As shown in Fig. 5.5, for each individual tested case, the result provided by our Top-Algorithm is almost same that provided by LINGO's global-solver, and the relative difference between our Top-Algorithm and LINGO's global-solver is less than 1%. The results in Fig. 5.5 validate the accuracy of our Top-Algorithm in solving Problem (TEM) despite its non-convexity.

Furthermore, to evaluate the performance of our proposed DRL-based algorithm in Figs. 5.6 and 5.7, we implement our online algorithm in Python with the TensorFlow-1.4.0 and Keras-2.1.3. We adopt a fully connected DNN consisting of

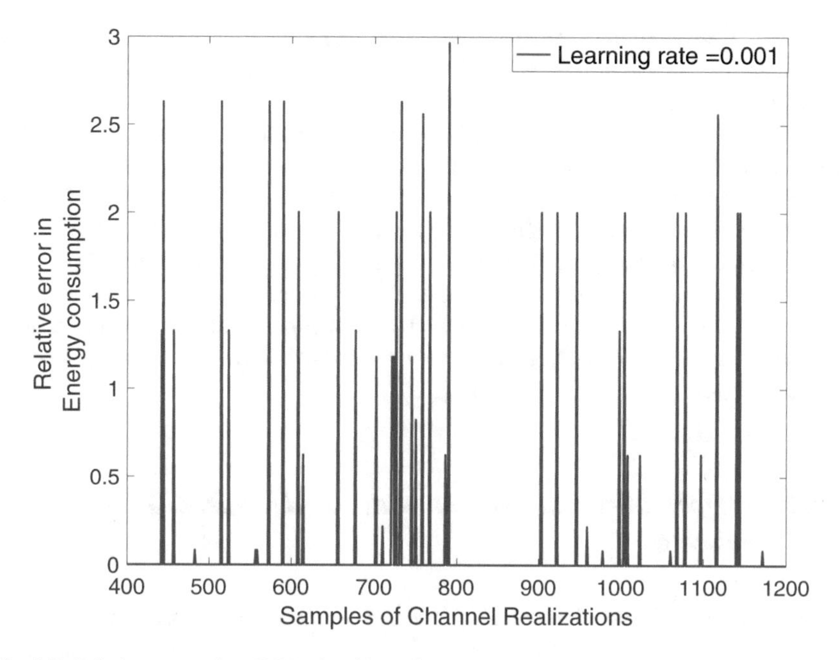

Fig. 5.7 Relative error of our DRL-algorithm after reaching stable

one input layer, two hidden layers, and one output layer. The two hidden layers have 1024 hidden neurons, in which the fully connections of neurons are dropped out with a probability of 0.3 when training. We set the training interval $N_{\mathrm{rt}} = 100$, the batch size for training as $|\mathcal{T}| = 100$, and the size of the training memory as 200. We generate 2000 different channel realizations as discussed at the beginning of Sect. 5.3 before.

Figure 5.6 illustrates the convergence of our proposed DRL-based online algorithm. Specifically, the top-subplot of Fig. 5.6 demonstrates the convergence of training loss when our DRL-based algorithm runs for 12 iterations of training-cycle, with each training-cycle covering 100 experienced samples over time. For the sake of evaluation, we test two cases of the learning rate, i.e., the learning rate of 0.01 and the learning rate of 0.001. Every point in Fig. 5.6 denotes the average result of 200 experienced samples over time. As shown in top-subplot of Fig. 5.6, for each tested case, the training loss of our DRL based algorithm quickly decreases when the learning process continues. These results indicate that our proposed DRL based algorithm can produce the solution very close to the optimal one after executing a sufficient amount of the rounds of pre-training. To further validate the results in top-subplot of Fig. 5.6, we also demonstrate the relative difference between the WD's energy consumption provided by our DRL-based algorithm and the exact minimum one for Top-Algorithm. The results are demonstrated in the bottom subplot of Fig. 5.6. We again test two cases of the learning rate, i.e., the learning rate of 0.01 and the learning rate of 0.001. As shown in the bottom subplot of Fig. 5.6, the relative error quickly decreases when the learning process continues, and in particular, the relative error is below 5% after around six training-cycles. This result is consistent with the result shown in the top subplot of Fig. 5.6.

Followed by Figs. 5.6 and 5.7 further shows the accuracy of our proposed DRL-based algorithm after the DNN module is trained to be stable. In particular, we compare the WD's energy consumption provided by our DRL-based algorithm with the exact minimum one provided by Top-Algorithm, and demonstrate the relative error in Fig. 5.7 for 800 samples of the channel realizations. As shown in Fig. 5.7, the relative error is almost negligible for most of the channel realizations, and the largest relative error is no greater than 3% for a very few number of channel realizations. The results in Fig. 5.7 demonstrate that our proposed DRL-algorithm can provide the near-optimal solution for Problem (TEM).

5.3.2 Performance Comparison with FDMA Based Offloading Schemes

In this subsection, we demonstrate the performance comparison between our proposed NOMA assisted multi-access offloading scheme in comparison with the conventional FDMA based offloading scheme. Specifically, in the FDMA-based scheme, we divide the channel bandwidth into several small pieces, with each piece of the channel bandwidth used for the WD to offload its workloads to one of the ECSs. Thus, with FDMA, there will be no co-channel interference among the WDs' offloading transmissions towards different ECSs. However, the downside of using FDMA is that the offloading throughput to the ECSs might degrade due to the smaller channel bandwidth after separating the channel bandwidth.

Figures 5.8 and 5.9 firstly demonstrate the comparison results for the static channel scenario in which we adopt our proposed Top-Algorithm (and its subroutine) for solving Problem (TEM) and obtaining the optimal offloading solution via NOMA transmission. In Fig. 5.8, we test two cases, i.e., the case of 3-ESs and 2 tasks and the case of 5-ESs and 2 tasks. For each tested case, we further vary S_k^{tot} from 6 Mbits to 11 Mbits. As shown in Fig. 5.8, it is reasonable to observe that the total energy consumption increases with respect to S_k^{tot}. Moreover, our proposed NOMA-based multi-task offloading scheme can always outperform the FDMA-based scheme. This result is reasonable, since leveraging NOMA can enable a full utilization of the channel bandwidth while using the SIC to mitigate the co-channel interference among the WDs' offloading transmission towards different ECSs and thus can yield a larger overall offloading throughput.

Followed by Fig. 5.8, we further vary $T_k^{\max}$ and compare the performance of our NOMA-based multi-task offloading scheme and the FDMA based offloading scheme. The corresponding results are shown in Fig. 5.9. As shown in Fig. 5.9, it

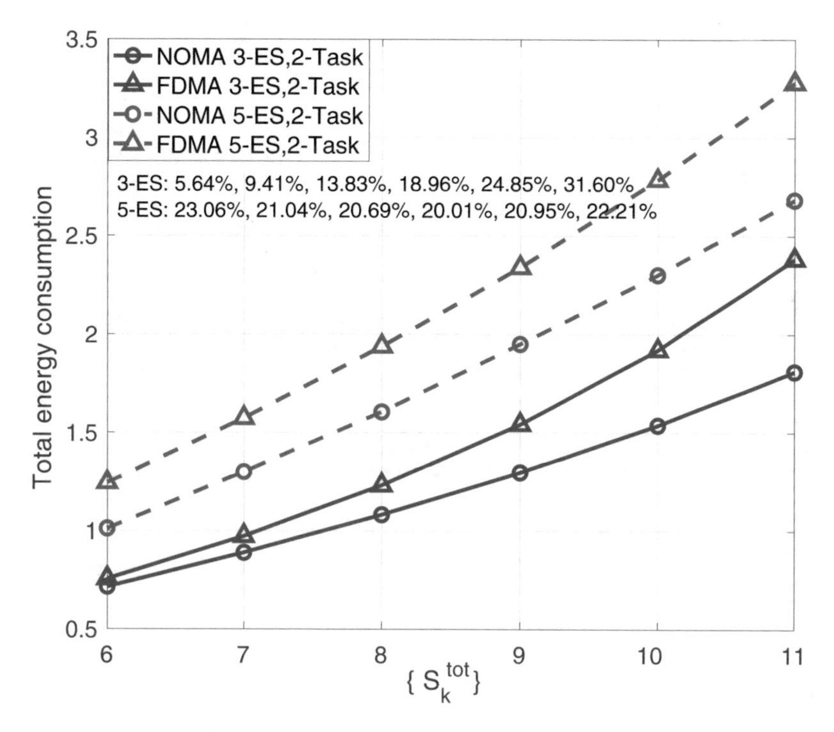

Fig. 5.8 Performance advantage against FDMA with $[S_1^{tot}, S_2^{tot}] = [6, 8]$

is reasonable to observe that the total energy consumption decreases with respect to T_k^{max}. Moreover, our proposed NOMA-based multi-task offloading scheme can still always outperform the FDMA-based scheme, since NOMA can avoid the orthogonal separation of the bandwidth resource, which thus improves the offloading throughput.

Finally, we demonstrate the advantage of our NOMA-based offloading under the dynamic channel scenario, by using our proposed DRL based online algorithm. For the purpose of comparison, we also show the result of the FDMA offloading under the dynamic channel scenario. The comparisons are shown in Fig. 5.10 which demonstrates the results for 800 samples of the channel realizations. In particular, for the sake of clear presentation, each point in Fig. 5.10 denotes the average of 10 samples. The results in Fig. 5.10 demonstrate that thanks to the proposed DRL based online algorithm, the NOMA-based offloading scheme can always outperform the FDMA-based offloading scheme even under the dynamic channel scenario.

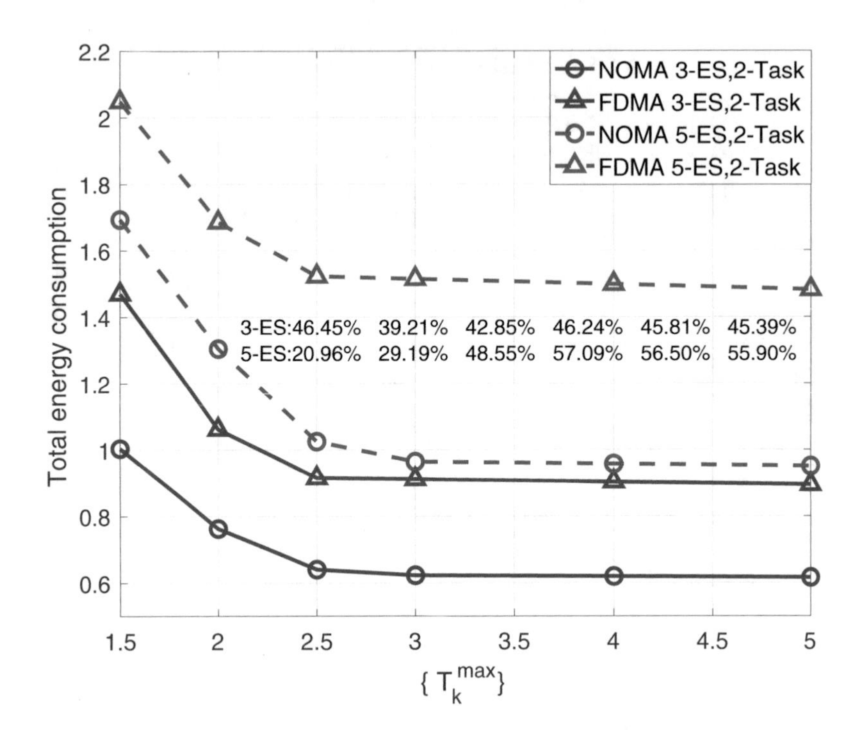

Fig. 5.9 Performance advantage against FDMA with $[T_1^{\max}, T_2^{\max}] = [2, 3]$

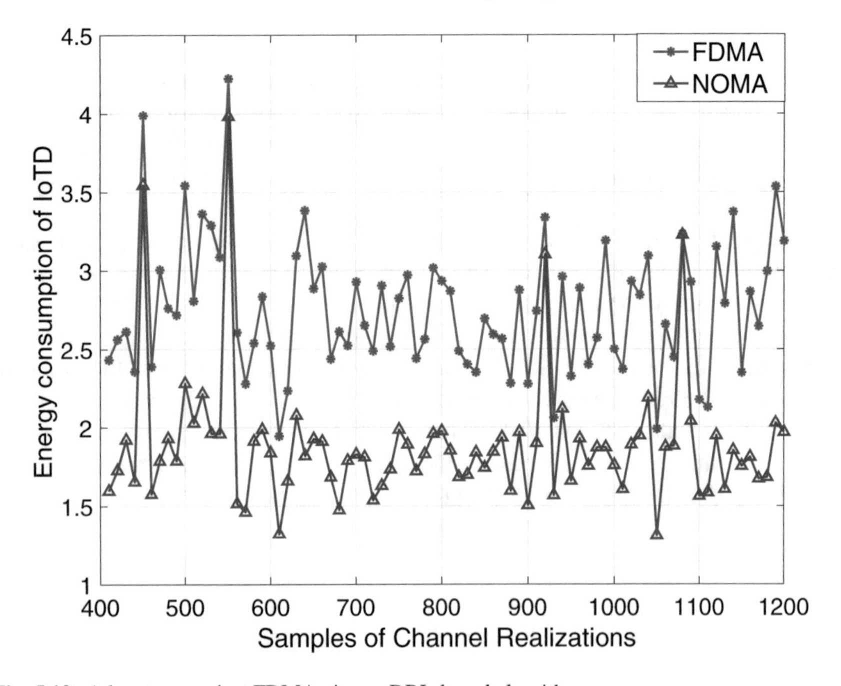

Fig. 5.10 Advantage against FDMA via our DRL-based algorithm

5.4 Literature Review

MA-MEC, which migrates cloud facilities to the edge of radio access network (RAN), has provided a promising approach to improve the quality of service of mobile devices when running computation-intensive yet delay-sensitive applications. Driven by its potential benefits, MA-MEC has received a lot of attention in recent years. In particular, there have been several survey studies focusing on MA-MEC. In [29], Abbas et al. provided an overview of MEC by focusing on the architecture, applications, resource management and security issue. The authors of [30] provided a comprehensive review of MA-MEC and focused on the forthcoming technologies integrated with MEC, including NOMA, wireless power transfer and energy harvesting, unmanned aerial vehicle (UAV), and heterogeneous cloud radio access network. Meanwhile, many research efforts have been devoted to different use-cases of MEC. In [31], Mao et al. surveyed the progress of MEC research from a communication perspective. In [32], Spinelli et al. provided a comprehensive overview of the relevant standards for MEC. In [33], Ranaweera et al. investigated the security and privacy issues in MEC as well as the possible solutions. In [34], Taleb et al. focused on the orchestration issues of different deployment schemes in MEC and the reference architectures. In [35], Mach et al. reviewed three key areas of the user-oriented computation offloading in MEC, namely, the decision on computation offloading, the allocation of computing resource, and the mobility management.

(Architecture of MEC and Multi-tier MEC): The reference framework and architecture of MEC are defined by the European Telecommunications Standards Institute. In [36], Sabella et al. further summarized and explained the framework and architecture of MEC. There have been many studies focusing on the MEC frameworks under different areas. In [37], Wang et al. proposed a framework of MA-MEC networks employing massive MIMO and small cells, which mitigates the interference, increases the throughput, and alleviates the backhaul capacity. In [38], to improve the quality and fairness of MEC services, Liu et al. proposed a network function parallelism MEC framework that outperforms conventional software defined network (SDN) enabled MEC. In [39], Zhao et al. proposed a three-level optimization framework to improve the performance of offloading computational tasks. In [40], Huang et al. proposed a SDN based MEC framework that integrates MEC into the third-generation partnership project (3GPP) architecture. The framework provides the required data-plane flexibility and programmability, and yields the latency improvement. To overcome the shortcomings of existing MEC frameworks in supporting Industrial Internet of Things (IIoT), in [41], Hou et al. proposed an MEC framework for IIoT by using Docker containers to slice the computation and storage resources of MEC servers. To support the latency requirements of growing number of connected IoT devices, the multi-tier edge computing framework has been studied in several papers [42, 43]. By exploiting the rich resources (e.g., communication, computation, and storage capabilities) of multi-tier servers, this framework enables cooperation among the servers deployed in different tiers. Thus, the multi-tier edge computing framework offers enhanced computing performance and provides more flexible computing services. Nevertheless, how to design an efficient cooperation strategy, i.e., communication mode, offloading decisions, and schedule algorithm, is still an open question. The work in [44] considers the joint optimization of transmission power and computational/radio resources allocation in a multi-tier edge-cloud system. In [45], the servers selection, transmit power allocations, and the received beamformers are jointly optimized in a heterogeneous cellular network. In addition, the authors in [46] focused on both the horizontal and vertical collaboration of edge servers. The advantages of MEC and MA-MEC have inspired lots of applications. In [47], Porambage et al. investigated the exploitation of MEC for IoT systems with various application scenarios. In [48], Jiang et al. provided a comprehensive summary of the current state of the art in applying MEC for video streaming systems. In [49], Siriwardhana et al. extensively discussed the future prospects of mobile augmented reality assisted by MEC. In [50], Khan et al. studied the advances in MEC-enabled smart cities. In [51], Amin et al. analyzed the MEC architectures and technologies for smart healthcare systems. In [52], Arthurs et al. surveyed the important advances of MEC applications in smart connected vehicles and intelligent transportation systems. In [53], Shi et al. investigated the MEC applications in agriculture systems and discussed the corresponding key architectures and technologies.

(Resource Allocation for MEC and Energy Efficiency): Resource management plays a crucial role in achieving the benefits of MA-MEC, which thus has attracted lots of research interests [2–4, 54–57]. In [2], Peng et al. exploited MA-MEC for enabling a SDN-based resource management for autonomous vehicular networks. In [3], a collaborative task offloading scheme has been proposed for vehicular edge multi-access networks. In [4], a joint optimization of computation offloading and transmission time allocation scheme has been proposed for multi-access offloading. In [54], Guo et al. focused on a fiber-wireless networks and proposed a collaborative multi-access computation offloading scheme. In [55], Huang et al. jointly optimized the video segment caching, transcoding and resource block assignment for adaptive video streaming for MA-MEC networks. In [56], Liang et al. tackled the challenge of user mobility in MA-MEC. In [57], Chai et al. designed an optimal strategy for optimizing the computation offloading and resource allocation in cellular device-to-device MEC systems.

With the growing concerns on green wireless systems in the past decades, many research efforts have been devoted to investigating energy-efficient MEC [6, 7, 26, 27, 58–61]. In [6], Zhou et al. investigated the energy-efficient workload offloading problem and proposed a distributed solution based on consensus alternating direction method. In [7], Chang et al. proposed a multi-objective optimization framework for computation offloading, which accounts for the balance between the energy consumption and delay performance. In practice, an IoT device may run several tasks simultaneously. In [26], Kuang et al. considered the multi-task computation offloading, and proposed a partial offloading scheduling and resource allocation scheme to minimize the weighted sum of the energy consumption and execution delay. In [27], the authors took task dependency into account and proposed a task offloading and resource allocation scheme to minimize the weighted sum of the energy consumption and task execution time. In [58], Malik et al. jointly optimized the resource allocation of data transmission, computing and wireless charging resources to minimize the weighted sum of energy consumption of all users and MEC servers. In [59], Chen et al. focused on the energy-efficient dynamic offloading in IoT MEC and used stochastic optimization to minimize the average transmission energy consumption while guaranteeing the average queue length for each device. In [60], Pu et al. studied a parallel MA-MEC system and investigated the energy-efficient resource management scheme. In [61], Eom et al. proposed an asynchronous design based offloading protocol that jointly optimizes the transmit power, the bandwidth allocation, the time duration, and task offloading partition, with the objective of minimizing the total energy consumption.

(Secrecy Driven for MEC): With the growing concerns on the secrecy of wireless services, secrecy driven MEC has attracted lots of interests in recent years [62]. In particular, the open access nature of radio channel has posed a particular security issue to MEC, namely, the malicious user can intentionally overhear the offloading transmissions of the edge-computing users by collecting the radio signals. To address the eavesdropping attack, physical layer security has been used to quantify how the secure the edge-computing user's offloading transmission is and the corresponding secrecy based offloading schemes have been proposed for MEC and Multi-access MEC [63, 64]. Recently, the covert communication, which aims at preventing an adversary from knowing whether the transmission is undergoing or not, has been considered as an approach for securing the offloading transmission [65]. To directly reduce the overhearing capacity of the malicious eavesdropper, artificial jamming has been leveraged for enhancing the secrecy of the offloading transmission. In [66], Bai et al. proposed a jamming scheme to secure the UAV's computation task offloading by considering both the active and passive eavesdropper. In [67], different from the previous studies, Wang et al. considered a dual-role UAV and leveraged the computation offloading transmission to provide artificial jamming for the relay transmission.

(Deep Learning based Designs for MEC): Thanks to the recent advances in artificial intelligence and deep learning (DL), there have been several studies exploiting DL for MEC [8–10, 12–14, 68–72]. For instance, in [8], Wang et al. exploited DRL for computing resource allocation and network resource allocation in mutative MEC environment for average service time minimization and resource allocation balancing. In [9], a framework that integrates DL based data validation and edge computing based local processing is proposed. In [10], the authors exploited the actor-critic DRL for jointly optimizing the caching, computing, and radio resource allocations in fog-enabled IoT systems. In [12], the authors proposed a model-free reinforcement learning offloading mechanism that learns the multi-user offloading strategies. In [69], He et al. developed a privacy-aware offloading scheme based on the deep post-decision state learning algorithm. In [70], Jiang et al. introduced deep neural networks with a scheduling layer to provide user association and computational resource allocation. In [71], Zhao et al. proposed a DRL based optimization framework for secure MEC networks. In [72], Shi et al. proposed a mean-field game guided DRL approach for task placement in cooperative MEC, which can help servers make timely task placement decisions to reduce the average service latency.

(NOMA assisted MEC): NOMA has been considered as one of the key technologies for accommodating massive connectivity in future wireless networks and enabling ultra-high spectrum-efficiency transmission [15, 16, 73]. Thanks to its great potential in enabling massive connectivity, many survey studies have been devoted to NOMA [74–77]. In [78], Ding et al. proposed a design of intelligent reflective surface assisted NOMA downlink transmission. In [79], Chen et al. applied NOMA to increase the total achievable rate of multi-user visible light communication systems. In [80], Mei et al. proposed a cooperative NOMA scheme by using the existing backhaul links between BSs for the cellular-connected UAV communications. In [81], Tang et al. investigated the joint power allocation and time-switching control for optimizing energy efficiency in a time-switching based simultaneous wireless information and power transfer NOMA system. In [82], Abu et al. discussed the problem of maximizing the total downlink rate of NOMA systems in the presence of imperfect SIC. In [83], Mu et al. studied a downlink multiple-input single-output intelligent reflecting surface assisted NOMA system and aimed at maximizing the total rate for all users. In [84], Qian et al. leveraged NOMA to provide cooperative jamming to the malicious eavesdropper and enhance the security of the offloading transmission of wireless devices. Thanks to the advantage of NOMA in enabling the highly efficient multi-access transmission, there have been many studies exploiting NOMA for MEC and MA-MEC [18–22, 24, 25, 85]. Due to the broadcasting nature of radio transmission, users' offloaded workloads to the MEC server over the wireless channel might be overheard by some malicious eavesdroppers. Thus, several studies have studied the security challenges in NOMA-MEC systems. In [86], Wu et al. studied the design of anti external eavesdropping in MEC network based on NOMA. To improve the security of offloaded tasks and enhance the users' connectivity, the authors jointly leveraged the physical layer security and NOMA for MEC networks.

(Future Directions of MEC): The past years have witnessed a deep penetration of MEC into various emerging applications. ETSI has intended to deliver interactive Virtual Reality, Augmented Reality and Mixed Reality over the emerging 5G/B5G networks and MEC sites. In [87], Mestres et al. combined software-defined networking, telemetry, network analytics and knowledge plane to propose a new paradigm, i.e., knowledge-defined networking, for intelligent prediction of offload costs and for efficient resource allocation of MEC servers. In addition, to enhance the security of MEC, blockchain and federated learning are two promising directions. In [88], Liu et al. proposed a secure data sharing scheme by using asynchronous learning methods in a blockchain-backed MEC system.

5.5 Summary

This chapter proposes the NOMA assisted multi-task MA-MEC via DRL. The chapter firstly considers a static channel scenario and formulates a joint optimization of the multi-access multi-task computation offloading, NOMA transmission, and computation-resource allocation, with the objective of minimizing the total energy consumption of wireless device to complete its tasks. Exploiting the layered structure of the problem, a layered algorithm is proposed to solve this non-convex optimization problem. Furthermore, the chapter investigates the time-varying channel scenario and proposes a DRL-based online algorithm to efficiently learn the near-optimal offloading solutions under time-varying channel realizations, which avoids solving the joint optimization problem for each channel realization.

There are several open future research directions for further extending the studies in this chapter.

- *(Multi-user multi-task offloading):* In this chapter, we mainly consider the multi-task offloading for a single WD. In practice, there exist a group of WDs executing their respective multi-task offloading simultaneously. Our study in this chapter can be further extended for investigating the multi-WD scenario. In particular, a hybrid NOMA-FDMA transmission scheme can be used for address multi-WD multi-tasks scenario, in which FDMA is used for enabling different WDs' simultaneous offloading transmissions with the allocated orthogonal subchannels, and NOMA is further used for each individual WD to offload its offloaded workloads to different ECSs simultaneously. Nevertheless, in this multi-WD scenario via hybrid NOMA-FDMA transmission scheme, we need to account for the channel bandwidth allocations for different WDs' FDMA transmissions in parallel. Moreover, we should take the computation-unit allocations of different ECSs for processing different WDs' offloaded workloads, e.g., to ensure the fairness among the WDs.
- *(Security in computation offloading):* This chapter considers an ideal scenario for the WD's computation offloading. In practice, there are some potential security issues in the computation offloading. Specifically, due to the open nature of the radio channel, a malicious user can intentionally collect signals of the WD's offloading transmissions and decode the task data in a brute-force manner. To address this challenge, the physical layer security can be leveraged for modeling how secure the WD's offloading transmission is. In particular, the physical layer security provides a fundamental measure of the secure throughput of the WD's transmission which cannot be overheard by any eavesdropper. Nevertheless, exploiting the physical layer security also leads to the additional resource allocation (e.g., additional energy consumption). Thus, it is an interesting direction for investigating the tradeoff between enhancing the security of the WD's offloading transmission and improving the utilization of computation/communication resource.
- *(UAV/drone assisted offloading):* In this chapter, we assume that the locations of the ECSs are all fixed and deployed on ground. Thanks to the recent

advances in UAV/drone networks, there has been a growing interest in exploiting UAV/drone networks for facilitating the computation offloading. In particular, thanks to the mobility of UAV/drone networks, the UAV/drone networks enable a promising approach for providing an *on-demand and flexible* platform that can accommodate the offloaded workloads from the WDs. The UAV/drone assisted edge computing has attracted various applications, e.g., the emergency edge computing services without an infrastructure network. Nevertheless, due to the limited energy/battery capacity of UAV/drone, the success of the UAV/drone assisted edge computing necessitates an energy management that jointly accounts for the energy consumption for providing edge-computing services, the energy consumption for the operations of UAV/drone (e.g., the mobility), and the energy consumption for communications. For instance, there have been several studies investigating the joint optimization of the trajectory and computation offloading for UAV enabled edge computing networks.

- *(Incentive Mechanism for multi-task multi-access offloading):* This chapter considers that the WD and all ESCs work together for improving the energy efficiency of the edge computing systems. The recent years have witnessed a growing interest in leveraging incentive mechanisms (e.g., auction mechanisms and contract designs) for resource allocations in wireless services and networks. By accounting for each agent's individual interest, the incentive mechanism can provide a promising approach for enabling a self-motivated yet distributed decision process, which thus avoid the requirement a central authority for executing the algorithms. Therefore, under the scenario of multi-access offloading provided by multiple ECSs, the WD can adopt the incentive mechanism such that different ECSs, in a self-motivated manner, are willing to accommodate different amounts of the offloaded computation workloads from the WD, depending on the ESCs' respective interests. This leads to an interesting direction to further extend the multi-access offloading in this chapter.

References

1. T. Taleb, K. Samdanis, B. Mada, H. Flinck, S. Dutta, D. Sabella, On multi-access edge computing: a survey of the emerging 5g network edge cloud architecture and orchestration. IEEE Commun. Surv. Tutor. **19**(3), 1657–1681 (2017). Thirdquarter
2. H. Peng, Q. Ye, X. Shen, Sdn-based resource management for autonomous vehicular networks: a multi-access edge computing approach. IEEE Wirel. Commun. **26**(4), 156–162 (2019)
3. G. Qiao, S. Leng, K. Zhang, Y. He, Collaborative task offloading in vehicular edge multi-access networks. IEEE Commun. Mag. **56**(8), 48–54 (2018)
4. Y. Wu, K. Ni, C. Zhang, L. Qian, D. Tsang, Noma-assisted multi-access mobile edge computing: a joint optimization of computation offloading and time allocation. IEEE Trans. Vehic. Technol. **67**(12), 12244–12258 (2018)
5. L. Qian, Y. Wu, N. Yu, F. Jiang, H. Zhou, T.Q. Quek, Learning driven noma assisted vehicular edge computing via underlay spectrum sharing. IEEE Trans. Vehic. Technol. **70**(1), 977–992 (2021)
6. Z. Zhou, J. Feng, Z. Chang, X. Shen, Energy-efficient edge computing service provisioning for vehicular networks: a consensus admm approach. IEEE Trans. Vehic. Technol. **68**(5), 5087–5099 (2019)

7. L. Liu, Z. Chang, X. Guo, S. Mao, T. Ristaniemi, Multiobjective optimization for computation offloading in fog computing. IEEE Internet Things J. **5**(1), 283–294 (2017)
8. J. Wang, L. Zhao, J. Liu, N. Kato, Smart resource allocation for mobile edge computing: a deep reinforcement learning approach. IEEE Trans. Emerg. Top. Comput. **9**(3), 1529–1541 (2019)
9. Z. Zhou, H. Liao, B. Gu, K. Huq, S. Mumtaz, J. Rodriguez, Robust mobile crowd sensing: When deep learning meets edge computing. IEEE Network **32**(4), 54–60 (2018)
10. Y. Wei, F. Yu, M. Song, Z. Han, Joint optimization of caching, computing, and radio resources for fog-enabled iot using natural actor-critic deep reinforcement learning. IEEE Internet Things J. **6**(2), 2061–2073 (2018)
11. Y. Chen, Z. Liu, Y. Zhang, Y. Wu, X. Chen, L. Zhao, Deep reinforcement learning-based dynamic resource management for mobile edge computing in industrial internet of things. IEEE Trans. Ind. Inf. **17**(7), 4925–4934 (2020)
12. T. Dinh, Q. La, T. Quek, H. Shin, Learning for computation offloading in mobile edge computing. IEEE Trans. Commun. **66**(12), 6353–6367 (2018)
13. D. Zeng, L. Gu, S. Pan, J. Cai, S. Guo, Resource management at the network edge: a deep reinforcement learning approach. IEEE Netw. **33**(3), 26–33 (2019)
14. L. Li, K. Ota, M. Dong, Deep learning for smart industry: efficient manufacture inspection system with fog computing. IEEE Trans. Ind. Inf. **14**(10), 4665–4673 (2018)
15. Y. Wu, L. Qian, H. Mao, X. Yang, X. Shen, Optimal power allocation and scheduling for non-orthogonal multiple access relay-assisted networks. IEEE Trans. Mob. Comput. **17**(11), 2591–2606 (2018)
16. H. Zhang, B. Di, Z. Chang, X. Liu, L. Song, Z. Han, Equilibrium problems with equilibrium constraints analysis for power control and user scheduling in noma networks. IEEE Trans. Vehic. Technol. **69**(5), 5467–5480 (2020)
17. Y. Wu, Y. Song, T. Wang, L. Qian, T.Q. Quek, Non-orthogonal multiple access assisted federated learning via wireless power transfer: a cost-efficient approach. IEEE Trans. Commun. (2022). https://doi.org/10.1109/TCOMM.2022.3153068
18. Z. Ding, D. Ng, R. Schober, H. Poor, Delay minimization for NOMA-MEC offloading. IEEE Signal Process. Lett. **25**(12), 1875–1879 (2018)
19. Q.-V. Pham, H. Nguyen, Z. Han, W.-J. Hwang, Coalitional games for computation offloading in noma-enabled multi-access edge computing. IEEE Trans. Vehic. Technol. **69**(2), 1982–1993 (2020)
20. A. Kiani, N. Ansari, Edge computing aware NOMA for 5G networks. IEEE Internet Things J. **5**(2), 1299–1306 (2018)
21. L. Qian, A. Feng, Y. Huang, Y. Wu, B. Ji, Z. Shi, Optimal SIC ordering and computation resource allocation in MEC-aware NOMA NB-IoT networks. IEEE Internet Things J. **6**(2), 2806–2816 (2019)
22. B. Liu, C. Liu, M. Peng, Resource allocation for energy-efficient mec in noma-enabled massive IoT networks. IEEE J. Select. Areas Commun. **39**(4), 1015–1027 (2021)
23. Y. Wu, L.P. Qian, K. Ni, C. Zhang, X. Shen, Delay-minimization nonorthogonal multiple access enabled multi-user mobile edge computation offloading. IEEE J. Select. Topics Signal Process. **13**(3), 392–407 (2019)
24. M. Sheng, Y. Dai, J. Liu, N. Cheng, X. Shen, Q. Yang, Delay-aware computation offloading in noma mec under differentiated uploading delay. IEEE Trans. Wirel. Commun. **19**(4), 2813–2826 (2020)
25. Y. Wu, B. Shi, L. Qian, F. Hou, L. Cai, X. Shen, Energy-efficient multi-task multi-access computation offloading via noma transmission for IoTs. IEEE Trans. Ind. Inf. **16**(7), 4811–4822 (2020)
26. Z. Kuang, L. Li, J. Gao, L. Zhao, A. Liu, Partial offloading scheduling and power allocation for mobile edge computing systems. IEEE Internet Things J. **6**(4), 6774–6785 (2019)
27. J. Yan, S. Bi, Y. Zhang, M. Tao, Optimal task offloading and resource allocation in mobile-edge computing with inter-user task dependency. IEEE Trans. Wirel. Commun. **19**(1), 235–250 (2019)

28. L. Qian, Y. Wu, F. Jiang, N. Yu, W. Lu, B. Lin, Noma assisted multi-task multi-access mobile edge computing via deep reinforcement learning for industrial internet of things. IEEE Trans. Ind. Inf. **17**(8), 5688–5698 (2020)
29. N. Abbas, Y. Zhang, A. Taherkordi, T. Skeie, Mobile edge computing: a survey. IEEE Internet Things J. **5**(1), 450–465 (2018)
30. Q.-V. Pham, F. Fang, V.N. Ha, M.J. Piran, M. Le, L.B. Le, W.-J. Hwang, Z. Ding, A survey of multi-access edge computing in 5G and beyond: fundamentals, technology integration, and state-of-the-art. IEEE Access **8**, 116974–117017 (2020)
31. Y. Mao, C. You, J. Zhang, K. Huang, K.B. Letaief, A survey on mobile edge computing: the communication perspective. IEEE Commun. Surv. Tutor. **19**(4), 2322–2358 (2017)
32. F. Spinelli, V. Mancuso, Toward enabled industrial verticals in 5G: a survey on MEC-based approaches to provisioning and flexibility. IEEE Commun. Surv. Tutor. **23**(1), 596–630 (2021)
33. P. Ranaweera, A.D. Jurcut, M. Liyanage, Survey on multi-access edge computing security and privacy. IEEE Commun. Surv. Tutor. **23**(2), 1078–1124 (2021)
34. T. Taleb, K. Samdanis, B. Mada, H. Flinck, S. Dutta, D. Sabella, On multi-access edge computing: a survey of the emerging 5G network edge cloud architecture and orchestration. IEEE Commun. Surv. Tutor. **19**(3), 1657–1681 (2017)
35. P. Mach, Z. Becvar, Mobile edge computing: a survey on architecture and computation offloading. IEEE Commun. Surv. Tutor. **19**(3), 1628–1656 (2017)
36. D. Sabella, A. Vaillant, P. Kuure, U. Rauschenbach, F. Giust, Mobile-edge computing architecture: the role of MEC in the internet of things. IEEE Consumer Electron. Mag. **5**(4), 84–91 (2016)
37. C. Wang, R.C. Elliott, D. Feng, W.A. Krzymien, S. Zhang, J. Melzer, A framework for MEC-enhanced small-cell hetnet with massive mimo. IEEE Wirel. Commun. **27**(4), 64–72 (2020)
38. M. Liu, G. Feng, Y. Sun, N. Chen, W. Tan, A network function parallelism-enabled mec framework for supporting low-latency services, in *IEEE Transactions on Services Computing* (2021), pp. 1–1
39. Z. Zhao, R. Zhao, J. Xia, X. Lei, D. Li, C. Yuen, L. Fan, A novel framework of three-hierarchical offloading optimization for MEC in industrial IoT networks. IEEE Trans. Ind. Inf. **16**(8), 5424–5434 (2020)
40. A. Huang, N. Nikaein, T. Stenbock, A. Ksentini, C. Bonnet, Low latency mec framework for SDN-based LTE/LTE-a networks, in *2017 IEEE International Conference on Communications (ICC)* (2017), pp. 1–6
41. X. Hou, Z. Ren, K. Yang, C. Chen, H. Zhang, Y. Xiao, IIoT-MEC: a novel mobile edge computing framework for 5G-enabled IIoT, in *2019 IEEE Wireless Communications and Networking Conference (WCNC)* (2019), pp. 1–7
42. J. Ren, D. Zhang, S. He, Y. Zhang, T. Li, A survey on end-edge-cloud orchestrated network computing paradigms: transparent computing, mobile edge computing, fog computing, and cloudlet. ACM Comput. Surv. (CSUR) **52**(6), 1–36 (2019)
43. D.E. Sarmiento, A. Lebre, L. Nussbaum, A. Chari, Decentralized SDN control plane for a distributed cloud-edge infrastructure: a survey. IEEE Commun. Surv. Tutor. **23**(1), 256–281 (2021)
44. E. El Haber, T.M. Nguyen, C. Assi, Joint optimization of computational cost and devices energy for task offloading in multi-tier edge-clouds. IEEE Trans. Commun. **67**(5), 3407–3421 (2019)
45. X. Hu, L. Wang, K.-K. Wong, M. Tao, Y. Zhang, Z. Zheng, Edge and central cloud computing: a perfect pairing for high energy efficiency and low-latency. IEEE Trans. Wirel. Commun. **19**(2), 1070–1083 (2019)
46. Y. Wang, X. Tao, X. Zhang, P. Zhang, Y.T. Hou, Cooperative task offloading in three-tier mobile computing networks: an ADMM framework. IEEE Trans. Vehic. Technol. **68**(3), 2763–2776 (2019)
47. P. Porambage, J. Okwuibe, M. Liyanage, M. Ylianttila, T. Taleb, Survey on multi-access edge computing for internet of things realization. IEEE Commun. Surv. Tutor. **20**(4), 2961–2991 (2018)

48. X. Jiang, F.R. Yu, T. Song, V.C.M. Leung, A survey on multi-access edge computing applied to video streaming: some research issues and challenges. IEEE Commun. Surv. Tutor. **23**(2), 871–903 (2021)
49. Y. Siriwardhana, P. Porambage, M. Liyanage, M. Ylianttila, A survey on mobile augmented reality with 5G mobile edge computing: architectures, applications, and technical aspects. IEEE Commun. Surv. Tutor. **23**(2), 1160–1192 (2021)
50. L.U. Khan, I. Yaqoob, N.H. Tran, S.M.A. Kazmi, T.N. Dang, C.S. Hong, Edge-computing-enabled smart cities: a comprehensive survey. IEEE Internet Things J. **7**(10), 10200–10232 (2020)
51. S.U. Amin, M.S. Hossain, Edge intelligence and internet of things in healthcare: a survey. IEEE Access **9**, 45–59 (2021)
52. P. Arthurs, L. Gillam, P. Krause, N. Wang, K. Halder, A. Mouzakitis, A taxonomy and survey of edge cloud computing for intelligent transportation systems and connected vehicles, in *IEEE Transactions on Intelligent Transportation Systems* (2021), pp. 1–16
53. X. Shi, X. An, Q. Zhao, H. Liu, L. Xia, X. Sun, Y. Guo, State-of-the-art internet of things in protected agriculture. Sensors **19**(8) (2019) [Online]. Available: https://www.mdpi.com/1424-8220/19/8/1833
54. H. Guo, J. Liu, Collaborative computation offloading for multiaccess edge computing over fiber-wireless networks. IEEE Trans. Vehic. Technol. **67**(5), 4514–4526 (2018)
55. X. Huang, L. He, L. Wang, F. Li, Towards 5G: joint optimization of video segment caching, transcoding and resource allocation for adaptive video streaming in a multi-access edge computing network. IEEE Trans. Vehic. Technol. **70**(10), 10909–10924 (2021)
56. Z. Liang, Y. Liu, T.-M. Lok, K. Huang, Multi-cell mobile edge computing: joint service migration and resource allocation. IEEE Trans. Wirel. Commun. **20**(9), 5898–5912 (2021)
57. R. Chai, J. Lin, M. Chen, Q. Chen, Task execution cost minimization-based joint computation offloading and resource allocation for cellular D2D MEC systems. IEEE Syst. J. **13**(4), 4110–4121 (2019)
58. R. Malik, M. Vu, Energy-efficient joint wireless charging and computation offloading in MEC systems. IEEE J. Select. Topics Signal Process. **15**(5), 1110–1126 (2021)
59. Y. Chen, N. Zhang, Y. Zhang, X. Chen, W. Wu, X. Shen, Energy efficient dynamic offloading in mobile edge computing for internet of things. IEEE Trans. Cloud Comput. **9**(3), 1050–1060 (2021)
60. X. Pu, W. Feng, W. Wen, Q. Chen, Energy-efficient parallel multi-access edge computing in OFDMA wireless networks. IEEE Trans. Vehic. Technol. **70**(9), 9613–9618 (2021)
61. S. Eom, H. Lee, J. Park, I. Lee, Asynchronous protocol designs for energy efficient mobile edge computing systems. IEEE Trans. Vehic. Technol. **70**(1), 1013–1018 (2021)
62. D. He, S. Chan, M. Guizani, Security in the internet of things supported by mobile edge computing. IEEE Commun. Mag. **56**(8), 56–61 (2018)
63. L. Qian, Y. Wu, N. Yu, D. Wang, F. Jiang, W. Jia, Energy-efficient multi-access mobile edge computing with secrecy provisioning. IEEE Trans. Mob. Comput. (2021). https://doi.org/10.1109/TMC.2021.3068902
64. Y. Wu, J. Shi, K. Ni, L. Qian, W. Zhu, Z. Shi, L. Meng, Secrecy-based delay-aware computation offloading via mobile edge computing for internet of things. IEEE Internet Things J **6**(3), 4201–4213 (2018)
65. N.T.T. Van, N.C. Luong, H.T. Nguyen, F. Shaohan, D. Niyato, D.I. Kim, Latency minimization in covert communication-enabled federated learning network. IEEE Trans. Vehic. Technol. **70**(12), 13447–13452 (2021)
66. T. Bai, J. Wang, Y. Ren, L. Hanzo, Energy-efficient computation offloading for secure uav-edge-computing systems. IEEE Trans. Vehic. Technol. **68**(6), 6074–6087 (2019)
67. T. Wang, Y. Li, Y. Wu, Energy-efficient uav assisted secure relay transmission via cooperative computation offloading. IEEE Trans. Green Commun. Network. **5**(4), 1669–1683 (2021)
68. L. Zhang, J. Wu, S. Mumtaz, J. Li, H. Gacanin, J. Rodrigues, Edge-to-edge cooperative artificial intelligence in smart cities with on-demand learning offloading, in *Proc. of IEEE GLOBECOM* (2019), pp. 1–6

69. X. He, R. Jin, H. Dai, Deep pds-learning for privacy-aware offloading in mec-enabled IoT. IEEE Internet Things J. **6**(3), 4547–4555 (2019)
70. F. Jiang, K. Wang, L. Dong, C. Pan, W. Xu, K. Yang, Deep-learning-based joint resource scheduling algorithms for hybrid MEC networks. IEEE Internet Things J. **7**(7), 6252–6265 (2020)
71. R. Zhao, J. Xia, Z. Zhao, S. Lai, L. Fan, D. Li, Green mec networks design under uav attack: a deep reinforcement learning approach. IEEE Trans. Green Commun. Network. **5**(3), 1248–1258 (2021)
72. D. Shi, H. Gao, L. Wang, M. Pan, Z. Han, H.V. Poor, Mean field game guided deep reinforcement learning for task placement in cooperative multiaccess edge computing. IEEE Internet Things J. **7**(10), 9330–9340 (2020)
73. Z. Chang, L. Lei, H. Zhang, T. Ristaniemi, S. Chatzinotas, B. Ottersten, Z. Han, Energy-efficient and secure resource allocation for multiple-antenna noma with wireless power transfer. IEEE Trans. Green Commun. Netw. **2**(4), 1059–1071 (2018)
74. S. Islam, M. Zeng, O.A. Dobre, Noma in 5G systems: exciting possibilities for enhancing spectral efficiency. Preprint. arXiv:1706.08215 (2017)
75. L. Dai, B. Wang, Z. Ding, Z. Wang, S. Chen, L. Hanzo, A survey of non-orthogonal multiple access for 5G. IEEE Commun. Surv. Tutor. **20**(3), 2294–2323 (2018)
76. B. Makki, K. Chitti, A. Behravan, M.-S. Alouini, survey of noma: current status and open research challenges. IEEE Open J. Commun. Soc. **1**, 179–189 (2020)
77. O. Maraqa, A.S. Rajasekaran, S. Al-Ahmadi, H. Yanikomeroglu, S.M. Sait, A survey of rate-optimal power domain noma with enabling technologies of future wireless networks. IEEE Commun. Surv. Tutor. **22**(4), 2192–2235 (2020)
78. Z. Ding, H. Vincent Poor, A simple design of IRS-NOMA transmission. IEEE Commun. Lett. **24**(5), 1119–1123 (2020)
79. C. Chen, W.-D. Zhong, H. Yang, P. Du, On the performance of mimo-noma-based visible light communication systems. IEEE Photon. Technol. Lett. **30**(4), 307–310 (2018)
80. W. Mei, R. Zhang, Uplink cooperative noma for cellular-connected UAV. IEEE J. Select. Topics Signal Process. **13**(3), 644–656 (2019)
81. J. Tang, J. Luo, M. Liu, D. K. C. So, E. Alsusa, G. Chen, K.-K. Wong, J.A. Chambers, Energy efficiency optimization for noma with swipt. IEEE J. Select. Top. Signal Process. **13**(3), 452–466 (2019)
82. I. Abu Mahady, E. Bedeer, S. Ikki, H. Yanikomeroglu, Sum-rate maximization of noma systems under imperfect successive interference cancellation. IEEE Commun. Lett. **23**(3), 474–477 (2019)
83. X. Mu, Y. Liu, L. Guo, J. Lin, N. Al-Dhahir, Exploiting intelligent reflecting surfaces in noma networks: joint beamforming optimization. IEEE Trans. Wirel. Commun. **19**(10), 6884–6898 (2020)
84. L. Qian, W. Wu, W. Lu, Y. Wu, B. Lin, T.Q. Quek, Secrecy based energy-efficient mobile edge computing via cooperative non-orthogonal multiple access transmission. IEEE Trans. Commun. **69**(7), 4659–4677 (2021)
85. X. Zhang, J. Zhang, J. Xiong, L. Zhou, J. Wei, Energy-efficient multi-UAV-enabled multiaccess edge computing incorporating NOMA. IEEE Internet Things J. **7**(6), 5613–5627 (2020)
86. W. Wu, F. Zhou, R.Q. Hu, B. Wang, Energy-efficient resource allocation for secure NOMA-enabled mobile edge computing networks. IEEE Trans. Commun. **68**(1), 493–505 (2020)
87. A. Mestres, A. Rodriguez-Natal, J. Carner, P. Barlet-Ros, E. Alarcón, M. Solé, V. Muntés-Mulero, D. Meyer, S. Barkai, M.J. Hibbett et al., Knowledge-defined networking. ACM SIGCOMM Comput. Commun. Rev. **47**(3), 2–10 (2017)
88. L. Liu, J. Feng, Q. Pei, C. Chen, Y. Ming, B. Shang, M. Dong, Blockchain-enabled secure data sharing scheme in mobile-edge computing: an asynchronous advantage actor–critic learning approach. IEEE Internet Things J. **8**(4), 2342–2353 (2021)

Chapter 6
Conclusion

6.1 Concluding Remarks

In this book, we discuss energy efficient computation offloading and resource allocation for MEC in depth. However, the introduction of MEC brings challenges under energy-constrained and dynamic conditions. Therefore, it is very important to design a strategy for energy efficient computation offloading and resource allocation. In this book, we discuss issues such as task offloading, channel allocation, frequency scaling, and resource scheduling in MEC. Specifically, we study the energy efficient dynamic offloading problem in MEC, MEC-based energy efficient computation offloading and frequency scaling problem, deep reinforcement learning-based computation offloading problem under multi-edge servers, and NOMA-based multi-task and multi-access energy efficient computation offloading problem. We believe that the computation offloading and energy management solutions presented in this book can provide some valuable references for the practical application of MEC. Based on the discussion and analysis in this book, we give the following summary.

- MEC is considered as a promising computing paradigm, which deploys computing resources at the network edge close to end devices. For terminal devices with delay-sensitive and computation-intensive requirements, the introduction of MEC can greatly improve the quality of user experience. Since the edge servers are deployed at the edge of the network, it can not only reduce the task offloading delay of the terminal device, but also reduce the computing burden of the terminal device. In order to adapt to the dynamic channel conditions and provide better services for users, it is necessary to design an appropriate computation offloading scheme. A good computation offloading scheme can effectively improve the overall performance of the system. When designing the computation offloading scheme, the specific requirements of the designed system

Y. Chen et al., *Energy Efficient Computation Offloading in Mobile Edge Computing*, Wireless Networks, https://doi.org/10.1007/978-3-031-16822-2_6

should also be considered. We need to choose an appropriate method to design according to different system architectures and different requirements.

- With the development of the IoT, more and more computation-intensive and delay-sensitive applications are running on mobile devices, greatly increasing the energy consumption of the device. This situation can be improved by offloading some tasks to the MEC server. However, there is competition for resources among users, which leads to a longer task delay and lowers the user experience quality. In order to avoid this situation and reduce the transmission energy consumption of mobile devices, we control the offloading of mobile devices. In order to minimize the transmission energy consumption of mobile devices, we design a dynamic energy efficient offloading algorithm combined with stochastic optimization techniques. The algorithm can achieve a balance between queue length and energy consumption by adjusting the value of parameter V. At the same time, we also analyze the performance of the algorithm through experiments. The results show that the algorithm can effectively reduce the transmission energy consumption of the device while ensuring the stability of the queue backlog.
- Jointly optimizing task allocation and CPU cycle frequency is an efficient computation offloading method. Mobile devices, limited by battery life and computing power, are not sufficient to support computing tasks for long periods of time. The amount of energy a mobile device consumes when processing computing tasks is closely related to its CPU cycle frequency. At the same time, the amount of energy consumed by the mobile device when offloading computing tasks is closely related to the number of tasks it offloads, and a large delay may occur when tasks are not allocated properly. In addition, the mobile device will generate a certain amount of tail power consumption after completing the offloading of tasks. Therefore, in order to analyze the performance of the system more comprehensively, we consider both the assignment of tasks and the decision of the CPU cycle frequency of mobile devices. Since the generation and arrival of tasks are dynamic, and the state of the channel is also dynamic, we design a dynamic offloading algorithm that combines task offloading and CPU cycle frequency scaling using stochastic optimization techniques. At the same time, we also analyze the performance of the algorithm through experiments. The results show that the algorithm can effectively reduce the average energy consumption of the system under the constraint of queuing delay.
- The non-orthogonal multiple access (NOMA) technology can realize that users share the resources of the wireless system, so as to meet the needs of users for low delay. Using NOMA technology for computation offloading in MEC can improve the efficiency of wireless resource utilization and improve system performance. To reduce delay to a greater extent, tasks generated by mobile devices can be offloaded to multiple edge servers for execution. Using NOMA technology can offload workloads to multiple edge servers simultaneously. However, mobile devices often generate multiple tasks simultaneously. To assign different tasks to appropriate edge servers for processing, we implement NOMA-based multi-task computation offloading using deep reinforcement learning. To minimize the energy consumption of mobile devices, we jointly optimize NOMA transmission,

multitasking computation offloading, and computation resource allocation. We propose a hierarchical algorithm. Furthermore, since the wireless channel is time-varying, we propose an online algorithm based on deep reinforcement learning. This algorithm can get the approximate optimal offloading solution. At the same time, we also analyze the performance of the algorithm through experiments. The results show that the algorithm can improve the performance of the system under time-varying channel conditions.

6.2 Future Directions

This book introduces energy efficient computation offloading and resource allocation based on MEC, and proposes corresponding computation offloading methods. However, the research on MEC-based computation offloading is not very comprehensive, and we can do further research on it. Next, we propose several future research directions.

Energy-Latency Tradeoff for Dynamic Offloading in Mobile Edge Computing Systems with Energy Harvesting Devices MEC has powerful computing power, which can accelerate the running of applications, thereby improving the quality of user experience. However, with the rapid development of IoT technology, the number of mobile devices is also increasing dramatically. As a result, the computation-intensive tasks generated by applications are also increasing dramatically, which also brings great challenges to MEC servers. Due to the limited amount of energy that IoT devices can store, computing for such a large number of tasks cannot be supported. To this end, there are many studies considering the use of renewable energy to provide energy support for MEC systems. The MEC system with energy harvesting device can greatly alleviate the problem of computation delay caused by insufficient energy [1, 2]. We can deploy an energy module for mobile devices to collect energy converted from green energy. This allows for more efficient computation offloading. However, the collection of green energy (such as solar, wind, etc.) is affected by conditions such as time and geography. When the energy stored by the energy module is too low, it may not be enough to sustain the computation offloading. Therefore, we also need to design a corresponding computation offloading and resource allocation scheme to achieve sustainable computation offloading. For this, we can enable grid power supply when the energy of the energy harvesting device is low. Therefore, it becomes very important to design a computation offloading strategy that can take into account both.

Unmanned-Aerial-Vehicle-Assisted (UAV-Assisted) Computation Offloading for MEC Systems Due to the limitations of computing power and battery capacity, mobile devices cannot process the generated computation-intensive or delay-sensitive tasks in a timely manner. At this point, the mobile device can offload tasks to the edge server for execution to relieve its own pressure. However, since the location of the MEC server is fixed, flexibility is lost. UAVs have emerged as a promising solution due to their mobility and computing and communication capabilities [3, 4]. UAV-assisted edge computing can provide better guarantees for some emergency edge computing services. However, the UAV also consumes a certain amount of energy in the process of working. In order to make the UAV-assisted MEC system better serve users, we can consider adding a small energy harvesting device to the UAV. At the same time, we will also design a computation offloading scheme for it. We will jointly optimize the UAV's trajectory, resource allocation and computation offloading.

References

1. H. Hu, Q. Wang, R.Q. Hu, H. Zhu, Mobility-aware offloading and resource allocation in a MEC-enabled IoT network with energy harvesting. IEEE Int. Things J. **8**(24), 17541–17556 (2021). https://doi.org/10.1109/JIOT.2021.3081983
2. G. Zhang, W. Zhang, Y. Cao, D. Li, L. Wang, Energy-delay tradeoff for dynamic offloading in mobile-edge computing system with energy harvesting devices. IEEE Trans. Ind. Inf. **14**(10), 4642–4655 (2018). https://doi.org/10.1109/TII.2018.2843365
3. Q. Hu, Y. Cai, G. Yu, Z. Qin, M. Zhao, G.Y. Li, Joint offloading and trajectory design for UAV-enabled mobile edge computing systems. IEEE Int. Things J. **6**(2), 1879–1892 (2019)
4. A. Gao, Y. Hu, W. Liang, Y. Lin, L. Li, X. Li, A QoE-oriented scheduling scheme for energy-efficient computation offloading in UAV cloud system. IEEE Access **7**, 68656–68668 (2019)

Printed in the United States
by Baker & Taylor Publisher Services